68th Annual Meeting of the Labor and Employment Relations Association and LERA@ASSA Meeting 2016

Minneapolis, Minnesota, USA
26 – 29 May 2016

San Francisco, California, USA
6 – 8 January 2016

Editor:

Ariel Avgar

ISBN: 978-1-5108-6758-1

Printed from e-media with permission by:

Curran Associates, Inc.
57 Morehouse Lane
Red Hook, NY 12571

Some format issues inherent in the e-media version may also appear in this print version.

Printed by Curran Associates, Inc. (2018)

Labor and Employment Relations Association (LERA)
University of Illinois at Urbana-Champaign
121 Labor and Employment Relations Bldg.,
504 E. Armory Ave.
Champaign IL 61820
USA

Phone: (217) 333-0072
Fax: (217) 265-5130

LERAoffice@illinois.edu

Additional copies of this publication are available from:

Curran Associates, Inc.
57 Morehouse Lane
Red Hook, NY 12571 USA
Phone: 845-758-0400
Fax: 845-758-2633
Email: curran@proceedings.com
Web: www.proceedings.com

LABOR AND EMPLOYMENT RELATIONS ASSOCIATION SERIES

Proceedings of the LERA 2016 Meetings

LERA@ASSA Meeting
January 3-5, 2016
San Francisco, California

LERA 68th Annual Meeting
May 26-29, 2106
Minneapolis, Minnesota

LABOR AND EMPLOYMENT RELATIONS

ASSOCIATION SERIES

Proceedings of the LERA 2016 Meetings

LERA@ASSA Meeting
January 6–8, 2016, San Francisco, CA
(in conjunction with ASSA/AEA)

and

LERA 68th Annual Meeting,
May 26–29, 2016, Minneapolis, MN

Ariel Avgar, Editor-in-Chief

LABOR AND EMPLOYMENT RELATIONS ASSOCIATION SERIES
(Formerly the Industrial Relations Research Association)

Proceedings of the Annual Meeting
Annual Research Volume
Perspectives on Work Magazine

Inquiries and other communications regarding membership, meetings, publications, and general affairs of the Association, as well as notice of address changes, should be addressed to the LERA office.

LABOR AND EMPLOYMENT RELATIONS ASSOCIATION
University of Illinois at Urbana-Champaign School of Labor and Employment Relations
121 Labor and Employment Relations Building, MC-504
504 East Armory Avenue, Champaign, IL 61820
Telephone: 217/333-0072 Fax: 217/265-5130 E-mail: LERAoffice@illinois.edu
Website: http://lera.memberclicks.net EPRN: http://www.employmentpolicy.org/

CONTENTS

IV.
LERA/ISA JOINT SESSION—THE EVOLVING HEALTHCARE LANDSCAPE: HOW EMPLOYEES, ORGANIZATIONS, AND INSTITUTIONS ARE ADAPTING AND INNOVATING—LERA RESEARCH VOLUME 2016

Timothy Vogus, Presiding

V.
LERA BEST POSTERS

Ariel Avgar and Ryan Lamare, Presiding

VI.
OCCCUPATIONAL REGULATION IN THE AGE OF UBER

Robert J. Thornton, Presiding

VII.

DISTRACTION OR CATALYST? TRADE AGREEMENTS AND LABOR STANDARDS

Anil Verma, Presiding

VIII.
$15 MINIMUM WAGES: WHAT CAN THE RESEARCH TELL US?

Ken Jacobs, Presiding

IX.
WORK CHALLENGES IN FIRST NATIONS COMMUNITIES

John W. Budd, Presiding

X.
LERA BEST PAPERS I

Aaron J. Sojourner, Presiding

XI.
LERA BEST PAPERS II

Kenneth May, Presiding

XII.
LERA ANNUAL REPORTS

I. Whatever Happened to Incomes Policy? In Honor of the Late Lloyd Ulman

Neocorporatism and Incomes Policy in Europe: Past, Present, and Future

Barry Eichengreen
University of California, Berkeley

There exists a standard narrative about the role of labor markets and incomes policy in post-World War II Europe, or so I like to think.[1] The question I ask in this paper is how much of that standard narrative still applies in the 21st century.

The standard narrative goes roughly as follows. The European economy grew rapidly for a quarter of a century after World War II because it possessed a set of labor market institutions, which were in turn embedded in a larger matrix of social and political institutions that were well suited to the economic imperatives of the day. The problem in Europe then, similar to the problem in China and other emerging markets now, was to capitalize on the scope for catch-up growth. Europe was well inside the technological frontier, defined by the modern mass-production, high-speed-throughput methods pioneered by the United States. But the continent also possessed the capabilities needed to catch up in the Abramovitzian sense.[2] It had a relatively high level of human capital accumulated via education and training. It, or at least its western half, had relatively strong rule of law and contract enforcement. It had long experience with the market.

In order to grow, Europe then, like emerging markets today, needed to maintain a relatively high level of investment, capital being the vehicle through which frontier technology was transferred from abroad. It needed wage moderation for the profitability of that investment and to ensure that firms had the resources to finance capital formation. It needed a high level of solidarity—similar levels of compensation and job security for workers in different sectors and activities so that all parties to the bargain could see that the benefits of growth were widely shared.[3]

But workers would be willing to defer gratification in this way—to restrain the push for higher wages—only if they were confident that owners and managers would plow their profits back into investment rather than paying them out as dividends and managerial salaries. Workers would restrain their wage demands and invest in the training—making for faster growth and higher incomes in the future only if employment security was sufficient to ensure them that they would still be on the job when those higher incomes materialized and, if they were not, that they would be compensated by the welfare state. Wage restraint would be attractive for workers in a particular sector, moreover, only if they knew that workers in other sectors would exercise similar restraint.

These commitment and coordination problems were solved by the institutions of the European mixed economy, a system known to social scientists as "neocorporatism." Labor market institutions and incomes policies were integral to the operation of that system. These institutions included strong trade unions engaged in solidaristic bargaining that limited pay differentials for different classes of workers and whose agreements extended to the non-unionized segment of the economy. They included works councils and the German system of codetermination, in which workers occupied seats on corporate supervisory boards, enabling them to monitor the investment, dividend, and compensation practices of their employers. It included a role for government, which promulgated investment-friendly tax codes while penalizing dividends and conspicuous consumption, supplementing private investment with the necessary public infrastructure, and coordinating investments across sectors of the economy through policies of

Author's address: 530 Evans Hall, Berkeley, CA 94720-3880

indicative planning. In this way, government cast a watchful eye over both unions and employer associations, strengthening their incentive to adhere to the terms of their bargain. The result was high levels of investment, favorable productivity growth, and a stable labor share of GDP.

This simple story is far too simple, of course. Why, for example, should Europe have gotten a set of institutions ideally suited to economic imperatives of the day? Here it is important to resist the obvious functionalist logic: that Europe developed this institutional configuration because it was the most efficient solution to the economic challenges at hand. More accurately, the continent's institutional matrix was a fortuitous inheritance from the past. Or to mix metaphors, it was born of a painful period of labor. Union movements, socialist parties, and Catholic organizations (both religious and political) developed over long periods of time, coming to espouse an ethic of solidarity and ideology of corporatism. The neocorporatist arrangements of the post-World War II period were rooted in these earlier developments, as well as in 1930s experiments such as the Saltsjobaden agreement in Sweden and Popular Front policies in France and in less savory experiences with corporatism under Hitler and Mussolini. They built on the expanded role of the state in organizing the economy during World War II. This inheritance was fortuitously well suited and therefore readily adapted to the circumstances of the post-World War II era.

Then there is a question of how much weight, in explaining European economic growth, should be attached to these institutions, and in particular to centralized and concerted bargaining that limited pay differentials and encouraged wage restraint, as opposed to other factors, such as favorable demography, an ample supply of underemployed rural labor, and a buoyant external environment. The answer to this second question is "Surely, we don't know." Fortunately, all that is needed for the present formulation is for the institutions of neocorporatism, in the labor market and elsewhere, to have played some role.

Then came the slowdown in output and productivity growth in the final quarter of the 20th century. While this may have reflected the draining of the pool of underemployed rural labor and a less favorable external environment, it also reflected the growing mismatch between Europe's socioeconomic institutions and its growth problem. As the economy approached the technological frontier, incentives to innovate became increasingly important. As technical skills became more integral to the production process, the pressure for significant skill premiums intensified, coming into conflict with European societies' tradition of solidarity. Where such premiums were not forthcoming, skilled workers grew more militant, and unskilled workers were pushed off into the service sector. As the pool of rural labor was drained and unemployment fell to low levels, voluntary wage restraint became harder, leading European governments to supplement it with some form of incomes policy. Rising capital mobility, meanwhile, attenuated the link between domestic wage restraint and domestic investment. The result, by the 1980s, was not just slower growth but also rising unemployment, accelerating inflation, and increasingly heavy public debt burden.

All of these developments required Europe to grope toward a different model. Writing in 1999, Torben Iversen and I described this as a model of "limited decentralization" (alternatively, we might have called it a system of "fragmented neocorporatism"). By allowing limited wage differentials, greater labor market decentralization and differentiation promised to accommodate the demand for skilled workers generated by post-Fordist technologies. Where more wage and employment flexibility was needed, it was achieved by introducing flexible, part-time, fixed-term contracts for new (generally younger) workers, while continuing to protect the status and security of established union members, a corollary of which was rising wage dispersion.

Since continuous consultation between peak associations and the government could no longer be relied on for wage restraint, it was essential to develop another anchor for inflation expectations. In particular, it was important to signal that monetary policy would be nonaccommodating so that it was understood that excessive wage demands meant additional unemployment and not inflation. The new anchor was found in adopting an exchange rate commitment and giving the Central Bank the independence to pursue it.[4] In this way, the seed was planted for Europe's fatal decision to create the euro.

So matters stood at the dawn of the new century. How, finally, I now want to ask, should this story be updated in light of subsequent events? A first important observation highlighted by subsequent experience is that a hard commitment to nonaccommodating monetary policies is an inadequate substitute for the strong unions and corporatist cooperation that prevailed in Europe in the third quarter of the 20th century—and equally for the decentralized labor markets of the United States where the discipline of competition delivers wage restraint. There may have been no question about the commitment of the European Central Bank to nonaccommodating policies, but this did not prevent Greece, Spain, Portugal, and other countries from pushing wages far above sustainable levels in the first decade of the euro.[5] The subsequent readjustment, made necessary by the crisis, has been excruciatingly painful, although it would have been somewhat less painful with better crisis-management policies.

Second, German experience post-1999 casts additional light on prospects for European labor markets. The shock to the German economy from reunification had enduring effects. Workers in the less productive eastern *lander*, finding themselves unable to compete with more productive workers in the German west, abandoned the countrywide industrial agreements that governed their wages and terms of employment, agreements that had had the effect of pricing them out of jobs. The share of workers covered by collective agreements fell sharply with their defection. Chronic problems of unemployment and inadequate competitiveness in the first post-reunification decade then led Germany to move further in the direction of a decentralized labor market, reducing worker protections and creating new low-paid, short-term "mini jobs."

The culmination of this process was the so-called Hartz II reforms of the Social Democratic government of Gerhard Schroder in 2003-2005.[6] Restrictions on temporary employment agencies, temporary contracts, and working hours were relaxed. The entitlement period for unemployment benefits was shortened to 12 months.[7] In an echo of the post-World War II settlement, unions agreed to restrain their wage demands in order to go for growth and obtain higher incomes in the future. Employers for their part augmented productive capacity, in capital-goods-producing sectors in particular. Union–employer collaboration continued to focus on apprenticeship and other on-the-job training, equipping the German labor force with post-Fordist skills. Public employment offices were reorganized to better match workers with openings. Unlike the period of strong solidarity after World War II, however, the result was also growing pay differentiation and inequality.[8] Aided and abetted by strong demand from China and other emerging markets for German transport equipment and capital goods, the result was a more efficient, competitive, and profitable export sector, a "new German economic miracle," as it was sometimes portrayed.

The ultimate test of this new model was the 2008-2009 downturn. Outcomes were relatively benign: unemployment in Germany rose only modestly (from 7.4% to 7.9% between 2008 and 2009, after which it resumed its fall). The explanation for this favorable headline outcome is disputed, however.[9] One view is that Germany had moved further than other European countries in the American direction, decentralizing the labor market and eliminating rigidities.[10] This allowed wages and hours to be adjusted, maintaining something approaching full employment. The other view, toward which I am more inclined, is that institutions of solidarity remained sufficiently strong that unions, firms, and government were able to agree on exceptional measures to maintain employment in the crisis. Unions and employers negotiated flexible agreements for work sharing, while the government introduced a subsidy for short-time working hours.[11] There was extensive cooperation among firms and their works councils. In these so-called "alliances for jobs," workers agreed to wage restraints or even, in exceptional cases, to wage cuts, in exchange for which they were offered job stability.[12]

Finally, strong growth in the pre-crisis period created a shortage of workers in a variety of skill categories—workers that firms sought to hoard by limiting layoffs in the downturn.[13] In this interpretation, Germany did not simply move in the direction of more flexible U.S.-style labor markets, although it did decentralize and otherwise move some way in that direction. In addition, it also built importantly on the legacy of neocorporatism.

It is controversial to argue that German labor market arrangements point the way forward and should be emulated by other European countries. Germany's policies of wage restraint added to the difficulties of its neighbors both in the run-up to the crisis when low wage inflation aggravated competitiveness problems in southern Europe, and then after the crisis struck and Germany refused to match wage cuts in southern Europe with wage increases at home as a corrective of intra-EU imbalances. Germany had special advantages and an unusually strong incentive to trade modest amounts of wage restraint and flexibility for strong employment growth as a result of a favorable external environment (strong Chinese demand for its machinery exports—something not also enjoyed by southern European economies).

That said, German experience since 2003 points to institutions and structures conducive to growth during expansions and to stability during contractions in this new 21st-century world. The postwar bargain of wage restraint in exchange for capacity expansion, in its 21st-century incarnation, makes for a strong supply side. Greater wage differentiation than in the third quarter of the 20th century makes the social solidarity supporting this bargain more difficult to maintain, but work-sharing arrangements can offset this, to some extent, by contributing to the sense of collective responsibility. An additional benefit of this greater wage differentiation is an increased supply of skilled workers and technicians, who are needed by both large and middle-sized German firms using post-Fordist technologies. The downside of the model is excessive reliance on short-term contracts, leading to the emergence of a two-tier labor market and creating problems that the country will have to rectify as it looks to the future.

It will not be easy for other European countries, lacking Germany's strong neocorporatist tradition and favorable product mix and with social compacts frayed by the pressure of the crisis, to emulate this example. But then, no one said it would be. And no one was right.

Endnotes

[1] That narrative as I describe it here draws on Eichengreen (2007) and Eichengreen and Iversen (1999), which in turn draw on the influential work of Flanagan, Soskice, and Ulman (1983).

[2] See Abramovitz (1986).

[3] And so that no one, therefore, had an incentive to renege on the agreement.

[4] The British case is proof by counterexample: it was the one European country where strong unions were broken; competitive pressure could therefore be relied on to produce restraint, and the anchor of a fixed exchange rate was therefore rejected.

[5] See Bourgeot (2013).

[6] Followed by Hartz III, which I do not distinguish in what follows.

[7] With the exception of older workers.

[8] See Card, Heining, and Kline (2013).

[9] A review of the debate is found in Moeller (2010).

[10] See, for example, Boysen-Hogrefe and Grolle (2010).

[11] So-called "working hours accounts," put in place before the crisis, were designed to allow employers to ask workers to work longer hours than provided for under their collective agreements in return for working shorter hours thereafter, satisfying employers' need for flexibility but also labor's desire for shorter hours. In the downturn, they were used to shorten hours and share the work while pointing to the prospect of longer work weeks once the economy recovered.

[12] The bargain was not unlike that in the Nordic countries, where it was known as "flexicurity." See Masden (2006) for an overview of their experience.

[13] There may have also been a tendency on the part of German employers to under-hire in the preceding expansion (doubting that it would persist), bequeathing a shortage of employed workers in the downturn (Burda and Hunt 2011).

References

Abramovitz, Moses. 1986. "Catching Up, Forging Ahead, and Falling Behind." *Journal of Economic History,* Vol. 46, pp. 385-406.

Bourgeot, Remi. 2013. "Labor Costs and Crisis Management in the Eurozone: A Reinterpretation of Divergences in Competitiveness." Foundacion Robert Schuman (September).

Boysen-Hogrefe, Jens, and Dominik Grolle. 2010. "The German Labour Market Miracle." *National Institute Economic Review,* Vol. 214, pp. R38-R50.

Burda, Michael, and Jennifer Hunt. 2011. "What Explains the German Labor Market Miracle in the Great Recession?" *Brookings Papers on Economic Activity,* Vol. 1, pp. 273-335.

Card, David, Jorg Heining, and Patrick Kline. 2013. "Workplace Heterogeneity and the Rise of West German Wage Inequality." *Quarterly Journal of Economics,* Vol. 128, pp. 967-1015.

Eichengreen, Barry. 2007. *The European Economy Since 1945: Coordinated Capitalism and Beyond.* Princeton: Princeton University Press.

Eichengreen, Barry, and Torben Iversen. 1999. "Institutions and Economic Performance: Evidence from the Labour Market." *Oxford Review of Economic Policy,* Vol. 15, pp. 121-38.

Flanagan, Robert, David Soskice, and Lloyd Ulman. 1983. *Unionism, Economic Stabilization, and Incomes Policies: European Experience.* Washington, D.C.: The Brookings Institution.

Masden, Per Kongwhoj. 2006. "Flexicurity: A New Perspective on Labor Markets and Welfare States in Europe." Background paper, DG EMPL Seminar on Flexicurity, Brussels: European Commission (May).

Moeller, Joachim. 2010. "The German Labor Market Response in the World Recession: Demystifying a Miracle." *Zeitschrift fur Arbeitsmarkt Forschung,* Vol. 42, pp. 325-36.

I. Whatever Happened to Incomes Policy? In Honor of the Late Lloyd Ulman

Whatever Happened to U.S. Incomes Policy?

Robert J. Flanagan
Graduate School of Business
Stanford University

In the decades following WWII, incomes policies—rules intended to guide the evolution of wages and prices over time—became a standard element of macroeconomic policy packages in most advanced countries. Viewed internationally, the United States was a latecomer to peacetime incomes policy and one of the few countries to introduce an incomes policy *before* inflation was a problem, hoping to forestall inflation as the economy moved out of a recession. There has been no mention of using incomes policy for this or any other U.S. policy objective over the past 33 years. This paper briefly reviews the nature of U.S. postwar incomes policies and explores the reasons for their short life in the macroeconomic policy arsenal.

The Kennedy/Johnson Administration Policy: The Wage-Price Guideposts

In the final pages of its January 1962 annual report, President Kennedy's Council of Economic Advisers (CEA) described a set of "guideposts for noninflationary wage and price behavior"—what turned out to be a stealth approach to an incomes policy for the United States (U.S. President 1962: 185-90). The guideposts were not announced as policy but rather were "suggested here as aids to public understanding" (U.S. President 1962: 186) and as standards by which the public could assess individual wage and price changes. No further monitoring mechanisms or compliance incentives accompanied the guideposts.

At the time, inflation was not an imminent threat: With the unemployment rate averaging 6.7%, the CPI increased by only 0.7% in 1961. But the CEA expressed the view that concentrations of power in labor and product markets could raise wages and prices before full employment was restored and hoped that the guideposts would thwart those increases as demand increased in labor and product markets. The Kennedy Administration CEA hoped the policy would improve what was then perceived as a long-term trade-off between inflation and unemployment.

In devising the wage–price guideposts, the CEA also strove to avoid a rigid set of rules that would interfere with the relative wage and price adjustments that induce the reallocation of resources from less productive to more productive employments. As a result, a set of exceptions accompanied the guideposts. The basic principles were quite simple: The average rate of compensation should increase at the average rate of productivity. Since adherence to this rule would produce no increase in unit labor costs, the general price guideline was zero. However, the exceptions permitted higher wage increases in sectors where it was difficult to attract labor or where workers had a weak bargaining position and lower wage increases in industries with relatively high unemployment or significant bargaining power. In product markets, prices could rise more rapidly than zero in industries that experienced increases in nonlabor costs or had trouble attracting capital. Conversely, they could rise more slowly than zero in where "the relation of productive capacity to full employment demand shows the desirability of an outflow of capital" (U.S. President 1962: 189).

In their next three annual reports, the CEA simply restated the guideposts and exceptions with their rationale. By 1965, the unemployment rate had dropped to 4.5%, and annual inflation rose to 1.9%. In their January 1966

Author's address: Stanford Graduate School of Business, 655 Knight Way, Stanford, CA 94305 report, the CEA acknowledged official attempts at persuasion ("jawboning") in individual situations and recommended a general wage guidepost of 3.2% a year—the trend rate of growth of productivity (U.S. President 1966: 88-92). As markets tightened, the exceptions received less and less emphasis in CEA annual reports.

The Nixon Administration Policy

The 1970 annual report of President Nixon's council offered the following summary of the Kennedy/Johnson experience with incomes policy:

> As originally put forth the guideposts were to serve a general educational function of encouraging voluntary patterns of behavior that would be noninflationary. There was no suggestion that the Government would apply them in particular cases or try to enforce them. But it was natural to question whether actions in particular cases conformed to the guideposts, and the Government felt it necessary to comment on the justification for these actions. Once this threshold had been crossed, the Government also became involved in attempting to insure compliance in particular cases where it was considered necessary. … With the upsurge of inflation and inflationary pressure after mid-1965, the difficulty of reconciling the guideposts with market forces became more intense. (U.S. President 1970: 24)

But the new administration confronted a significant annual inflation rate (5.6% in 1970) and faced a different policy challenge: How to *unwind an existing inflation* that in some degree persisted because of expectations of future inflation. President Nixon responded in August 1971 with a 90-day freeze on prices, rents, wages, and salaries followed by a mandatory system of wage and price regulations.

Discussing the policy in their January 1973 report, the CEA claimed the policy had been successful but added: "Nevertheless, by the end of the year it was plain that although continuing controls could make a further contribution to economic stabilization, the system would have to be modified." (U.S. President 1973: 51). By now the program confronted significant supply shocks—initially food price increases and later oil price increases resulting from actions of the OPEC cartel. In its 1974 annual report, the CEA stated that "… the effect of the controls program on the rate of inflation in 1973 cannot be known with certainty either today or ever." (U.S. President 1974: 108)

A year later, with an inflation rate of 12.3% and unemployment at 5.6%, the CEA was even more specific.

> [A]lthough inflation might have been more rapid in the absence of controls, in the light of the actual experience both before and after their termination, it is difficult to accept that thesis. The lack of widespread support for extension of the controls program, not only in Congress but among business and labor representatives and the general public, suggests broad acceptance of this judgment. … [T]he net benefit of the controls system, however evaluated, had become extremely small by the beginning of 1974. …" (U.S. President 1975: 229)

In 1976, for the first time since the 1962 annual report, there was no discussion of incomes policy.

The Carter Administration Policy

The Carter Administration also faced the problem of how to unwind an existing and at times accelerating inflation and included incomes policy in its arsenal of anti-inflation weapons. Concerned with the limited achievements of incomes policies at home and abroad (Ulman and Flanagan 1971), several U.S. economists inside and outside of government noted that most incomes policies lacked strong incentives to comply. As a remedy, they proposed using the tax system to reward compliance and/or punish noncompliance and designed various tax-incentive plans (TIPs), which subsequently were reviewed by the Carter CEA (U.S. President 1981: 60-68.)

The Carter Administration initially requested each company to keep wage and price increases below the average for the prior two years. When that program proved ineffective, however, the administration proposed an incomes policy with specific maximum targets for wage and price increases, accompanied by a proposal for "real wage insurance" intended to insure workers against losses in real and relative wages that could occur because of the noncompliance of other workers. Congress never acted on the real wage insurance proposal, and TIP incentives disappeared from the landscape of U.S. incomes policy. Compliance rested on the force of public opinion and a threat that federal government contracts would be denied to noncomplying organizations. The policy continued to encounter significant supply-side shocks that it was never designed to cope with—from increase in the prices of food and energy to a decline in productivity. By the end of the Carter Administration, the CEA estimated that the policy had reduced annual pay and price increases by 1% to 1.5% in 1979 (U.S. President 1980: 36).

Incomes policy completely disappeared from U.S. macroeconomic policy with the arrival of the Reagan Administration, whose CEA bluntly asserted that prior incomes policies had not stopped inflation, and "[i]nflation is essentially a monetary phenomenon. … [P]ersistent inflation can[not] be explained by nonmonetary factors." (U.S. President 1982: 54). Over the subsequent 34 years, incomes policy has not reappeared on the American scene. This development cannot be attributed solely to the macroeconomic perspective of the Reagan CEA, for neither subsequent Democratic nor Republican administrations have proposed such policies. The rest of this paper considers why incomes policy fell from favor.

Accounting for the Demise: Underlying Theory

The use of incomes policies to thwart *prospective* inflation lacked a secure theoretical foundation. The motivation for the guideposts, for example, stressed that concentrations of economic power could lead to wage and price decisions that were not in the public interest. But the degree of monopoly power in labor or product markets produces a once-and-for-all increase in wages or prices—not ongoing inflationary pressure—unless those increases are accommodated by monetary policy. Absent Central Bank accommodation, only ongoing increases in market power could, by themselves, produce continual upward pressure on wages and prices.

In the face of declining unionization, however, it simply became increasingly difficult to connect pay inflation to the exercise of bargaining power. Union strength (measured by membership) in the United States had peaked in the mid-1950s—several years before the announcement of the wage-price guideposts. When the guideposts were announced in the early 1960s, private sector union density had already begun a long decline from roughly 30% of private wage and salary employment to 6.6% in 2014 (U.S. Bureau of Labor Statistics 2015). Competition from the growing non-union sector and foreign countries limited both the ability of unions to dominate pay setting and the ability of large companies to pass cost increases on into prices to the extent that an alleged connection between collective bargaining and general pay movements during the most recent U.S. recovery would have been fanciful.

In the 1962 guideposts, the CEA also expressed concern with restraining possible pay spillovers from the union to the non-union sector. Pattern bargaining had been an important feature of wage determination within the unionized sector during the early postwar period, but in 1962, little was known about the extent of spillovers to the non-union sector, which if accommodated by monetary policy could produce upward pressure on pay and prices. Later research found no evidence of significant spillovers (Flanagan 1976). The main growth sector of union organization, the public sector, had no significant bargaining over pay when U.S. incomes policies were in their heyday and do not set patterns for private sector wages today. In short, the connection between bargaining power and inflation had weak theoretical roots and little empirical traction.

A second theoretical difficulty with early U.S. incomes policies was the belief that the Phillips curve could be shifted. The distinction between the short-run and long-run Phillips curves was not clarified until the late 1960s and was not widely accepted for several more years. As the notion of an equilibrium rate of unemployment (Friedman's "natural rate" or in modern parlance, the NAIRU—nonaccelerating inflation rate of unemployment) became accepted, the role of incomes policy in restraining outbreaks of inflation became less obvious. Any rate of inflation was consistent with the NAIRU; the actual inflation rate rested on the expectations of future inflation generated by a country's history. But with expectations increasingly driven by foreign supply decisions (notably OPEC) rather than the behavior of domestic institutions, the role for a U.S. incomes policy to moderate inflationary expectations became less and less convincing.

Accounting for the Demise: Evidence of Effectiveness

Finally, U.S. experience with incomes policies provided little basis for returning to them. As we have seen, various CEAs struggled to find convincing evidence of their effectiveness, and all who tried to administer the policies came away impressed by the difficulties encountered in translating seemingly simple rules for wage and price behavior into practice in the thousands of union contracts and company human resource management systems that condition a modern labor market. Doubts presented by the U.S. evidence were reinforced by the lack of demonstrable long-term influence in the many versions of incomes policies tried abroad (Flanagan, Soskice, and Ulman 1983).

To this point, one must add the fact that (with the exception of the early 1960s) inflation has been lower since the demise of U.S. incomes policies (see Table 1, which shows inflation and unemployment rates by presidential administration). Good fortune played a role—the food and energy inflation record shows the diminished role of supply shocks since 1980, for example. But politicians and voters may have concluded that the classic tools of fiscal and monetary policies can produce a more reliable containment of inflation than incomes policies.

TABLE 1
Inflation and Unemployment Rates by Presidential Term

	Annual Average Percent Change in CPI				
Term	All items	All minus food & energy	Food	Energy	Unemployment Rate
1967-64	1.2	1.4	2.0	0.8	5.8
1965-68	3.3	3.4	3.3	1.7	3.9
1969-72	4.9	4.7	4.6	3.4	5.0
1973-76	8.2	7.2	9.8	14.3	6.7
1977-80	10.4	9.6	10.1	17.6	6.5
1981-84	5.1	5.9	3.5	3.2	8.6
1985-88	3.4	4.2	3.8	-9.2	6.5
1989-92	4.2	4.3	3.6	4.4	6.3
1993-96	2.8	2.8	3.5	2.0	6.0
1997-2000	2.4	2.3	2.1	3.8	4.4
2001-04	2.3	2.0	2.6	8.3	5.5
2005-08	2.5	2.2	3.8	4.0	5.0
2009-12	2.2	1.7	1.9	8.2	9.0

Source: Bureau of Labor Statistics.

In short, while the goals and mechanisms of peacetime incomes policies in the United States changed over time, none of the formulations produced lasting results. This evidence—along with a moderation of inflation in recent decades—reduced the political incentive to take direct regulatory action to address wage and price movements and limited these policies to a 20-year run in the United States.

References

Flanagan, Robert J. 1976. "Wage Interdependence in Unionized Labor Markets." *Brookings Papers on Economic Activity*, pp. 635-673.

Flanagan, Robert J., David W. Soskice, and Lloyd Ulman. 1983. *Unionism, Economic Stabilization and Incomes Policies: European Experience*. Washington, DC: Brookings Institution.

Ulman, Lloyd, and Robert J. Flanagan. 1971. *Wage Restraint: A Study of Incomes Policies in Western Europe*. Berkeley: University of California Press.

U.S. Bureau of Labor Statistics. 2015: www.bls.gov/news.release/union2.toc.htm

U.S. President. 1962. *Economic Report of the President and The Annual Report of the Council of Economic Advisers* (ERP hereafter). Washington, DC: GPO.

U.S. President. 1966. *ERP*. Washington, DC: GPO.
U.S. President. 1970. *ERP*. Washington, DC: GPO.
U.S. President. 1973. *ERP*. Washington, DC: GPO.
U.S. President. 1974. *ERP*. Washington, DC: GPO.
U.S. President. 1975. *ERP*. Washington, DC: GPO
U.S. President. 1980. *ERP*. Washington, DC: GPO.
U.S. President. 1981. *ERP*. Washington, DC: GPO.
U.S. President. 1982. *ERP*. Washington, DC: GPO.

I. Whatever Happened to Incomes Policy? In Honor of the Late Lloyd Ulman

Incomes Policy in Germany and Partial Decentralization of Collective Bargaining

Knut Gerlach
Institut für Arbeitsökonomik
Leibniz Universität Hannover

Introduction

This paper first analyzes the decline of collective bargaining and the onset of decentralized wage bargaining. Second, it explores the political and economic factors for establishment of an official incomes policy, its principal objectives, and the economic approaches that differed between unions and employers' associations—as well as its ultimate failure. Third, it is emphasized that after the demise of official incomes policy, the decentralization of wage determination proceeded, and flexible wage bargaining at the level of establishment—incorporating elements of the former official incomes policy—strongly expanded. Empirical results of this institutional turning point, including severe shortcoming, are highlighted.

Decline in Collective Bargaining and the Onset of Decentralization

Subject to some data restrictions, trends are presented for periods from the last decade of the 2000s through the second decade of the 21st century.

Net union density (not counting retiree members) of the DGB unions (unions affiliated with the German Federation of Unions) declined from 27.3% (1980) to 17.2% (2000) to 12.9% (2011). Union density in Germany was one of the lowest of the EU countries in 2010 (Ebbinghaus and Göbel 2014: 216, 230).

A similar trend is detectable in the German Employers' Associations (BDA and subgroups), especially with small and medium-sized firms withdrawing from or not joining the associations. In the Metal Employers' Associations of West Germany, for example, the percentage of membership firms declined from approximately 56 (1984) to 41 (1994) to 18 (2011) (Schroeder and Sylvia 2014: 354).

The coverage of sectoral bargaining fell from 68% (43%) of all private sector employees in West (East) Germany in 1996 to 60% (39%) in 2000 to 47% (29%) in 2014 (Ellguth and Kohaut 2015: 293).

The sectoral spread of collective bargaining coverage in 2010 varies widely, ranging from 69% in chemicals to 15% in IT services. Twenty percent of workers in small companies (i.e., companies with 10-49 employees) are covered by collective contracts, in contrast with 86% of workers in companies with more than 1,000 employees (Statistisches Bundesamt 2013). Average wages are highest in firms with sectoral contracts, followed by companies with an orientation toward sectoral collective wages and finally by their nonoriented counterparts (Addison 2012).

From 1996 to 2014, works councils existed in about 10% of firms, with no detectable trend. However, the percentage of all employees in firms with a works council and a sectoral collective agreement in West (East) Germany declined from 41% (29%) in 1996 to 37% (25%) in 2000 to 28% (15%) in 2014 (Ellguth and Kohaut 2015: 296). The percentages are substantially higher for companies with more than 500 employees (Ellguth 2004: 166).

Author's address: Wirtschaftswissenschaftliche Fakultät, Institut für Arbeitsökonomik, Königsworther Platz 1, 30167 Hannover, Germany

The basic trends of an erosion of collective bargaining were already quite striking before 1998, the year of the introduction of the second version of incomes policy in Germany, and the erosion proceeded unabatedly after the turn of the century.

In addition, this development was accompanied by steps toward a partial decentralization of collective bargaining. The high and persistent unemployment in West Germany the metal workers' union (IGM) in 1984 was accompanied by a long national strike for a reduction of weekly working time from 40 to 35 hours with full income compensation in order to redistribute working hours across more employees and thus reduce unemployment (Giersch, Paqué, and Schmieding 1992: 215). Finally, a compromise incorporated two elements: a gradual introduction of a working week of 35 hours and "opening clauses for working time" (OC–WT) for flexibility at the company level to be negotiated by the employer and the works council. After 1992, in a very severe recession, large companies negotiated cost-cutting measures with works councils and the consent of trade unions; these measures included steps such as working-time reductions without wage compensation, and employment guarantees (Rehder 2003: 116).

In this period, the government and the two bargaining partners agreed to transfer the basic labor market institutions to the eastern part of the country, and the unions specified rapid annual wage increases to attain western wages. Unemployment reached unsustainable levels, and the gap between wages and labor productivity increased. Employers demanded "opening clauses for employment and competitiveness" (OC–EC) in collective contracts in the east that would allow *firms, with the consent of their employees or the works council,* to pay wages below the standards of the collective wage agreement, increasing the likelihood of the survival of firms and stabilizing employment. Reluctantly, the unions gave in, fearing that this new type of opening clause might spill over to the west.

Second Version of Incomes Policy: Alliance for Employment, Training, and Competitiveness (1998–2003)

Twenty-one years after the termination of the Concerted Action (1967-1977), which brought the first version of an incomes policy in Germany (Ulman and Flanagan 1971; Flanagan, Soskice, and Ulman 1983), a second version was established by the newly elected coalition government of the Social Democratic and Green Parties (Center/Green) in a period of high rates of unemployment. Several factors might have contributed to this decision.

First, against the background of stagnating and declining employment, IGM and DGB launched an initiative in 1995 for an improvement in employment. The issues were discussed in a forum for nonbinding discussions among the government, unions, and employers' associations (Bispinck and Schulten 2000: 6). After the Center/Right coalition enacted fiscal spending cuts in the area of social policy, the unions terminated their participation.

Second, the idea of a social pact at the national level was adopted by the Social Democratic Party and was strongly supported by unions in the election campaign.

Third, the major components of the German system of industrial relations—works councils (co-determination), collective agreements, unions, and employers' associations have been declining. The decline, however, is much less pronounced in large companies. These relations are characterized by stable union density rates and high rates of coverage by collective agreements and works councils, and they try to use wage cuts by reducing voluntary bonuses (Hassel 1999: 502). Because large companies play important roles in employers' associations, they have an incentive to participate in an alliance that might lead to more moderate wage hikes and reduced unemployment.

The alliance was chaired by the chancellor and consisted of the major business organizations and most important sectoral unions. After the first meeting of the alliance, a joint consensual statement declared the reduction of high unemployment as the most serious challenge requiring permanent cooperation among the state, unions, and employers (Arlt and Nehls 1999: 262-264). Subsequently, unions (DGB) and employers' associations agreed on the preferred use of productivity increases for employment expansion, the reduction of working time for additional jobs, and an extension of working-time accounts (OC–WTs).

Despite this compromise, the main focuses of unions and employers' associations differed. Unions generally preferred working-time reductions in order to redistribute the available volume of work without income cuts and/or an extension of early retirement, while the employers' associations opted strongly for wage settlements below productivity increases. However, in the bargaining round for 2000 and 2001, the chemical workers union (IGBCE), referring explicitly to the alliance's employment-oriented collective bargaining policy, took the wage-leader function and settled for moderate wage increases. The public reaction to the chemical agreement was very positive; it influenced the negotiations of IGM and led to moderate wage increases in that sector (Bispinck and Schulten 2000: 20).

During this bargaining round, unemployment declined and wages rose moderately. With the onset of recession in 2002, the still high unemployment rate stagnated and then began to rise again. The results of collective bargaining were criticized by IGM members, who began a series of short strikes and initiated a harsh debate about the Alliance and the participation of unions in the tripartite institution. The ensuing contractual wage increases for 2003 and 2004 were generally regarded as excessive. Apparently, for the strategy and actions of German unions, fairness and equity are more important tenets than a reduction of unemployment (Ulman, Gerlach, and Giuliano 2005: 9).

With collective bargaining policy as a taboo topic for unions, a gradual phasing out of the alliance was unavoidable. After the reelection of the Center/Green coalition in 2002, the chancellor dissolved the alliance and abrogated the second German attempt for an official incomes policy.

Decentralization of Wage Determination

The main indicators of collective bargaining—net union density, membership in employers' associations, coverage by sectoral collective wage agreements, the proportion of employees in firms with a works council, and a sectoral collective wage agreement—continued to decline after the turn of the century, as discussed previously. The decline was very severe in East Germany and in general did not affect large companies.

After the demise of the alliance and with high and rising unemployment rates, "opening clauses for employment and competitiveness" (OC–EC) were established in West Germany. In a very important treaty negotiated for the engineering industry in 2004, IGM and the employers' association agreed on regulated firm-level bargaining between management and the works council with initially stringent supervisory rights of unions and employers' associations. On the one hand, this treaty debilitated the institutional governance of the labor market by the traditional bargaining partners, exposing it more strongly to market forces. On the other hand, and especially in the core of the German economy, the bargaining partners—management and works councils—reach a compromise: workers accepted pay cuts, extended working time and more flexible working arrangements, while management promised investments, guaranteed production at the location of the firm and employment, and abstained from dismissals.

In 2011, 28% of employees with a collective contract in the private sector were employed in establishments with an OC–WT, and 16% had an OC–EC. The percentages increased with plant size, and the two versions of these arrangements coexisted in many firms (Ellguth and Kohaut 2014: 442).

In the first decade of the 21st century, six independent occupational unions were founded with about 200,000 members (Lesch 2008: 146). These occupational unions organized qualified employees and competed with unions of the DGB. Their threats and frequent strikes were very successful in attaining higher wages, counteracting the flat wage structure in collective wage agreements (Schroeder, Kalass, and Greef 2011: 92). The debut of occupational unions led to an additional decentralization and fragmentation of the German system of collective wage contracts.

Reforms of the supply side of the labor market accelerated fragmentation and the partial erosion of collective bargaining. The basic proposals of the Hartz Commission were implemented between 2003 and 2005, directly after and as a response to the failure of the alliance. The Hartz activation policy tightened the rules for the unemployed to accept job offers. The Hartz welfare reforms cut the duration of unemployment insurance benefits (UB I). The merger of social assistance with assistance for long-term unemployment into a single flat rate and means-tested benefit generated a general minimum income with strong activation requests (UB II). These policy changes augmented labor supply for low-wage jobs.

In the course of the Hartz reforms, various types of atypical employment were deregulated. Fixed-term contracts were generally available to firms for up to two years. Agency (temporary) work was completely deregulated. In the core economy especially, works councils and management welcomed the flexibility provided by agency workers because they supported job stability for core workers. Finally, the Hartz reforms adopted, modified, and improved the appeal of marginal employment (mini-jobs).

In summary, these developments led to an increasing fragmentation and heterogeneity of the labor market and the labor force. Wage setting at the individual level became much more widespread. Due to the increasing availability and use of opening clauses, the role of firm-level works councils in bargaining was strengthened relative to unions (Carlin, Hassel, Martin, and Soskice 2015: 83; Dustmann, Fitzenberger, Schönberg, and Spitz-Oener 2014). The activation policy for labor supply and the partial deregulation of atypical forms of labor contributed to the acceptance of low-wage jobs. In addition, the fiscal deficit rules of the "Stability and Growth Pact" in the eurozone limit the application of fiscal policy, whereas the monetary policy of the ECB regulates the inflation rate in the entire eurozone, not just in Germany. While these institutional changes further weakened the case for official incomes policies, they simultaneously supported the transfer and imitation of some of their elements, such as bargaining over wages and employment, to the microeconomic level.

Flexible Lower-Level Bargaining: Empirical Results

The prior argument raises two issues. First, why is the core of the German economy—which is composed of larger and middle-sized firms in manufacturing, banking, finance, insurance, and energy with sectoral collective bargaining and works councils, although benefitting from partial decentralization—reluctant to advocate a cancellation of the system of sectoral collective bargaining (Thelen 2000)? Freeman and Lazear (1995: 49) show that "works councils are most likely to improve enterprise surplus when they have limited but definite power in the enterprise." This encourages works councils to focus first and foremost on improving the operation of the workplace, on enhancing productivity, and on reducing resignations and dismissals. A high proportion of the wage bundle is determined by sectoral collective contracts, and larger firms, in accordance with works councils, usually raise the remuneration. In periods requiring employment stabilization and an improvement of competitiveness, firms can still use sectoral collective wages as a benchmark and bargain with works councils over reductions of extra pay and within the framework of opening clauses of undercutting the benchmark. Empirical results for Germany support this analysis. The impact of works councils on wages is less strong in covered plants than in noncovered ones, and their productivity-enhancing effects are more likely to be seen in establishments with a collective wage contract (Hübler and Jirjahn 2003).

The second issue concerns the economic effects of opening clauses. OC–WTs are a compromise between management and works councils. They increase productivity and lead to a reduced fluctuation of workers. Wages and profits do not fall (Bellmann and Hübler 2015).

Slightly more than 50% of OC–ECs are used in economically critical phases of an establishment, whereas the remaining OC–ECs are intended to strengthen the future competitiveness of their plants starting from their still satisfactory economic condition (Hübler 2005, 2006a, 2006b; Bellmann, Gerlach, and Meyer 2008). The probability of a stable or rising employment is on average significantly higher in plants introducing these agreements during an acceptable economic condition with the goal of improving competitiveness than in establishments in an already critical phase. Trusting relationships between the bargaining partners are a significant ingredient of success.

In addition, a recent study of the effects of OC–ECs on employment in the severe economic crisis of 2008-2009 shows that in establishments affected by the crisis, the existence of an OC–EC supported the stabilization of employment (Bellmann and Gerner 2012).

In summary and with some caveats, the current empirical results show that opening clauses stabilize and stimulate employment, sometimes with wage moderation and particularly in larger companies.

The new institutional setup is correlated with an extreme increase of wage inequality. Germany now belongs to the group with the most pronounced earnings inequality in Europe (Rhein 2013). The increase of real mean wages was very modest in the period after 1995, whereas wage inequality rose very strongly. Real wages at the 15th percentile declined dramatically, at the median they started to fall after 2000, and at the 85 percentile, they continued to rise after 1995. The decline of union coverage and the decentralization of collective agreements as important elements of the remarkable change of wage inequality, however, vary between the studies (Antonczyk, Fitzenberger, and Sommerfeld 2010; Card, Heining, and Kline 2013; Dustmann, Ludsteck, and Schönberg 2009).

Conclusion

After 2004, with modestly rising real wages, the development of employment and labor productivity was more advantageous in Germany than in most other countries of the eurozone (Dustmann, Fitzenberger, Schönberg, and Spitz-Oehner 2014: 170; Thimann 2015: 145). The ongoing decentralization of wage determination and flexible lower-level bargaining were the principal driving forces of this process. A large part of total employment is relegated to competitive labor markets. Opening clauses (OCs) are important elements of the new focus of collective bargaining, mimicking to some degree past incomes policies. The case for official incomes policy became much weaker. However, the stability and future prospects of this development are at risk because works councils tend to foster the stabilization and growth of employment in conjunction with a sectoral collective wage contract. This side condition is weakening.

This paper is intended to show the significant role of works councils in the resurgence of the German economy. Works councilors are not necessarily union members, although many are. They have a margin of independence from the unions and often act for the benefits of firms. Because workers generally desire more voice and influence in firms and the workplace (Addison 2014: 8), these insights might be of some importance for the United States.

Acknowledgments

I am very grateful to Robert Flanagan, Olaf Hübler, and Wolfgang Meyer for valuable comments.

References

Addison, John T. 2012. "Is the Erosion Theory Overblown? Evidence from the Orientation of Uncovered Employers." *IZA Discussion Paper,* no. 6658, June 2012.

Addison, John T. 2014. "The Consequences of Trade Union Power Erosion." *IZA World of Work*, May 2014.

Antonczyk, Dirk, Bernd Fitzenberger, and Katrin Sommerfeld. 2010. "Rising Wage Inequality, the Decline of Collective Bargaining, and the Gender Wage Gap." *Labour Economics, Vol.* 17, pp. 835-847.

Arlt, Hans-Jürgen, and Sabine Nehls. 1999. *Bündnis für Arbeit.* Opladen/Wiesbaden: Westdeutscher Verlag.

Bellmann, Lutz, Knut Gerlach, and Wolfgang Meyer. 2008. "Company-Level Pacts for Employment." *Jahrbücher für Nationalökonomie und Statistik, Vol.* 228, no. 5+6, pp. 533-553.

Bellmann, Lutz, and Hans-Dieter Gerner. 2012. "Company-Level Pacts for Employment in the Global Crisis 2008/2009: First Evidence from Representative German Establishment-Level Panel Data." *The International Journal of Human Resource Management, Vol.* 23, no. 16, pp. 3375-3396.

Bellmann, Lutz, and Olaf Hübler. 2015. "Working Time Accounts and Performance in German Establishments." Unpublished Paper, Faculty of Economics, Leibniz Universität Hannover.

Bispinck, Reinhard, and Thorsten Schulten. 2000. "Alliance for Jobs—Is Germany following the Path of 'Competitive Corporatism'?" *WSI Discussion Paper,* no. 84 (Published by the Institute of Economic and Social Research of the Hans Böckler Foundation, Düsseldorf).

Card, David, Jörg Heining, and Patrick Kline. 2013. "Workplace Heterogeneity and the Rise of German Wage Inequality." *Quarterly Journal of Economics, Vol.* 128, no. 3, pp. 967-1015.

Carlin, Wendy, Anke Hassel, Andrew Martin, and David W. Soskice. 2015. "The Transformation of the German Social Model." In Jon Erik Dolvik and Andrew Martin, ed., *European Social Models from Crisis to Crisis—Employment and Inequality in the Era of Monetary Integration.* Oxford: Oxford University Press, pp. 49-104.

Dustmann, Christian, Bernd Fitzenberger, Uta Schönberg, and Alexandra Spitz-Oehner. 2014. "From Sick Man of Europe to Economic Superstar: Germany's Resurgent Economy." *Journal of Economic Perspectives, Vol.* 28, no. 1, pp. 167-188.

Dustmann, Christian, Johannes Ludsteck, and Uta Schönberg. 2009. "Revisiting the German Wage Structure." *Quarterly Journal of Economics, Vol.* 124, no. 2, pp. 843-881.

Ebbinghaus, Bernhard, and Claudia Göbel. 2014. "Mitgliederrückgang und Organisationsstrategien deutscher Gewerkschaften." In Wolfgang Schroeder, ed., *Handbuch Gewerkschaften in Deutschland*. Wiesbaden: Springer VS, pp. 207-239.

Ellguth, Peter. 2004. "Erosion auf allen Ebenen?" In Ingrid Artus, and Rainer Trinczek, ed., *Über Arbeit, Interessen und andere Dinge*. München and Mehring: Rainer Hampp Verlag, pp., 159-179.

Ellguth, Peter, and Susanne Kohaut. 2014. "Öffnungsklauseln—Instrument zur Krisenbewältigung oder Steigerung der Wettbewerbsfähigkeit?". *WSI Mitteilungen, Vol.* 67, no. 6, pp. 439-449.

Ellguth, Peter, and Susanne Kohaut. 2015. "Tarifbindung und betriebliche Interessenvertretung: Ergebnisse aus dem IAB-Betriebspanel 2014." *WSI Mitteilungen, Vol.* 68, no. 4, pp. 290-297.

Flanagan, Robert J., David W. Soskice, and Lloyd Ulman. 1983. *Unionism, Economic Stabilization, and Incomes Policies: European Experience*. Washington D.C.: Brookings Institution.

Freeman, Richard B., and Edward P. Lazear. 1995. "An Economic Analysis of Works Councils." In Joel Rogers and Wolfgang Streeck, ed., *Works Councils: Consultation, Representation and Cooperation in Industrial Relations*. Chicago: University of Chicago Press, pp. 27-50.

Giersch, Herbert, Karl-Heinz Paqué, and Holger Schmieding. 1992. *The Fading Miracle. Four Decades of Market Economy in Germany*. Cambridge: Cambridge University Press.

Hassel, Anke. 1999. "The Erosion of the German System of Industrial Relations." *British Journal of Industrial Relations, Vol.* 37, no. 3, pp. 483-505.

Hübler, Olaf. 2005. "Sind betriebliche Bündnisse für Arbeit erfolgreich? " *Jahrbücher für Nationalökonomie und Statistik, Vol.* 225, no. 6, pp. 630-652.

Hübler, Olaf. 2006a. "Quo Vadis, betriebliches Bündnis?" *Wirtschaftsdienst, Vol.* 86, no.2, pp. 96-101.

Hübler, Olaf. 2006b. "Zum Einfluss betrieblicher Bündnisse auf die wirtschaftliche Lage von Unternehmen." *Jahrbuch für Wirtschaftswissenschaften, Vol.* 57, pp. 121-146.

Hübler, Olaf, and Uwe Jirjahn. 2003. "Works Councils and Collective Bargaining in Germany: The Impact on Productivity and Wages." *Scottish Journal of Political Economy, Vol.* 50, no. 4, pp. 471-191.

Lesch, Hagen. 2008. "Spartengewerkschaften—Droht eine Destabilisierung des Flächentarifvertrags?" *Sozialer Fortschritt, Vol.* 57, no. 6, pp. 144-153.

Rehder, Britta. 2003. *Betriebliche Bündnisse für Arbeit—Mitbestimmung und Flächentarif im Wandel.* Frankfurt/New York: Campus.

Rhein, Thomas. 2013. "Deutsche Geringverdiener im europäischen Vergleich." *IAB-Kurzbericht,* no. 15, pp.1-10.

Schroeder, Wolfgang, and Stephen J. Sylvia. 2014. "Gewerkschaften und Arbeitgeberverbände." In Wolfgang. Schroeder, ed., *Handbuch Gewerkschaften in Deutschland*. Wiesbaden: Springer VS, pp. 337-365.

Schroeder, Wolfgang, Viktoria Kalass, and Samuel Greef. 2011. *Berufsgewerkschaften in der Offensive. Vom Wandel des deutschen Gewerkschaftssystem*. Wiesbaden: Springer VS.

Statistisches Bundesamt. 2013. *Tarifbindung in Deutschland 2010*. Reihe: Verdienste und Arbeitskosten. Wiesbaden.

Thelen, Kathleen. 2000. "Why German Employers cannot bring themselves to dismantle the German System." In Torben Iversen, Jonas Pontusson, and David W. Soskice, ed., *Unions, Employers, and Central Banks*. Cambridge: Cambridge University Press, pp. 138-169.

Thimann, Christian. 2015. "The Microeconomic Dimensions of the Eurozone Crisis and Why European Politics Cannot Solve Them." *Journal of Economic Perspectives, Vol.* 29, no. 3, pp. 141-164.

Ulman, Lloyd, and Robert J. Flanagan. 1971. Wage Restraint: A Study of Incomes Policies in Western Europe. Berkeley: University of California Press.

Ulman, Lloyd, Knut Gerlach, and Paola Giuliano. 2005. "Wage Moderation and Rising Unemployment—A Report on Union Moderation. " In Lutz Bellmann, Olaf Hübler, Wolfgang Meyer, and Gesine Stephan, ed., *Institutionen, Löhne und Beschäftigung*. Nürnberg: Beiträge zur Arbeitsmarkt- und Berufsforschung, no. 294, pp. 1-10.

State-Sponsored Unionization: The Development of Enterprise Unions in China

JIE WANG

National University of Singapore

The existing research asserts that enterprise unions are inherently weak due to their dependence on enterprise and state. This paper suggests that this structure-centric view cannot explain the diversity of enterprise unions at the firm level. This paper argues that union efficacy is determined by an enterprise's perception of state-sponsored unionization and a union chairperson's ability to make a union work. When enterprises perceive state-sponsored unionization as a resource, enterprise unions that are led by capable union leaders are more likely to develop specialized union functions and engage in effective collective action that promotes employees' economic interests. This paper is based on eight months of fieldwork in the southern China city of Shenzhen, where more than 50 interviews were conducted with enterprise unions, official unions, labor NGOs, and workers.

Introduction

Since the 1980s, when neoliberalism began to erode union movement in industrial democracies, unionism in East Asia has been on the rise. In South Korea, early entry into industrialization and the presence of an urban working class propelled independent union movements to join a prodemocracy resistance; this pressured the government to accept democratic transition. In countries such as China and Vietnam, which embarked on industrialization in the late 1970s and 1980s, the state not only leads industrialization, it continuously molds the formation and development of industrial relations. In these countries with a Leninist political tradition, union organizing is an important social and political task for the ruling parties because state-led unionization is employed by both party-states as a means to curb instability, in the form of either institutionalized labor disputes such as arbitrations and litigations or labor agitations such as wildcat strikes.

Focusing on China's state-sponsored unionization, this research explores the diversity of unionization outcomes in China. Transitioning from the early reform era (1970s-2000s) to the post-reform era (2008 to present), the state has become increasingly aware of the urgent need to balance growth and equity. Since the 2000s, a series of laws and supplementary regulations that address labor relations have been enacted to promote workplace stability and equity, to reduce labor-capital conflict, and to bring socially combustible labor protests into formal institutions that are promoted by the state. Two landmark legislations were passed during the 2000s: the Trade Union Law in 2001 and the Labor Contract Law in 2008. Both acts seek to increase the predictability of employment relations in favor of employees. The 2001 Trade Union Law, in particular, aims to enhance firm-level labor institutions to nullify labor disputes the workplace. This law expects enterprise unions to become a representative institution of employees and the primary conciliator of labor disputes at the firm level. This law also obligates enterprise unions to integrate the interests of employees and enterprises. By restricting union organizing at the firm level, the state attempts to simultaneously control labor organizations and monitor industrial relations at workplaces. The implementation of this law was assisted by the state-sponsored unionization soon afterward. Primarily targeting the private sector, the campaign that was implemented by official unions has significantly increased the number of enterprise unions. As Figure 1 demonstrates, the number of grassroots unions, mostly enterprise unions, had reached 2.6 million by 2012.

Author's address: Department of Political Science, National University of Singapore, AS1, #04-10, 11 Arts Link, Singapore 117570

However, a quantitative leap in the number of unions is far from indicative of the strength of organized labor because enterprise unions are registered by enterprises. Some scholars thereby suggest that the unionization campaign is merely a matter of formalism (Liu 2009), which serves to fulfill unionization quotas that are set by the official unions. However, it is clear that consent from enterprises is not possible if the state does not concede to them. Enterprises determine how unions operate and function to a large extent, despite intermittent guidance from official unions. Knowledge about how trade unions grow within this institutional constraint will complement the current research, which focuses primarily on unionism in developed countries. This research demonstrates how enterprises and the state interact to shape enterprise unions and the agent that they dispatch in order to make enterprise unions become what they expect to be.

FIGURE 1
Number of Grassroots Trade Unions (10,000 Units)

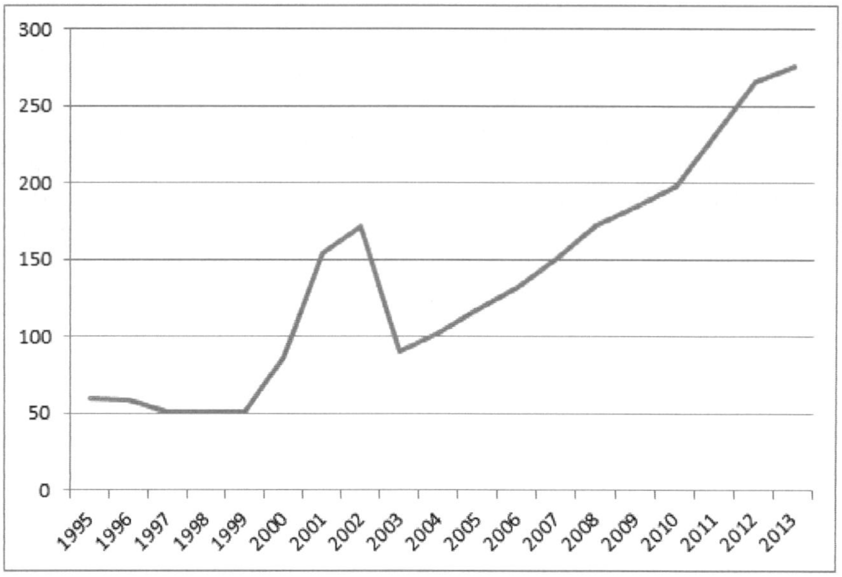

Literature Review

The research on unionism in China is derived from a structure-deterministic point of view, which perceives enterprise unions as dominated either by the state or the capital. The existing research can be divided into three primary approaches: union as a state instrument, union marginalization, and dual cooptation. The union as a state instrument approach is applied to explain unionism under state socialism. The union marginalization approach suggests that the market reform and the decline of the public sector led to an irreversible trend of union decline and marginalization. Finally, the dual cooptation approach argues that state-sponsored unionization leads to a cooptation of enterprise union by the capital and the state, which drives a hollowing-out of labor institution at the firm level.

Union as a State Instrument

Inspired by literature on totalitarianism, this approach suggests that the dominance of the Leninist party-state diminishes union autonomy. The Leninist state intends to maintain close control of social organizations and garner their obedience and support in the fulfillment of its collective social tasks (Linz 2000; Lenin 1921, cited in Feng 2006). In return for labor loyalty, state workers were guaranteed lifelong employment. With the absence of class conflict, the role of enterprise union was reduced to promote productivity, counter bureaucratism (Lenin 1921, cited in Feng 2006), and convey employees' opinions to the enterprise.

The dependence of unions on the state led to the decline of unions. Since the 1990s, the party-state has embarked on the reform of the public sector, which subjects state-owned enterprises (SOEs) to market pressure in order to spur the efficiency of this ill-performing sector. Furthermore, SOE managers have been given full authority to discipline and dismiss workers (Lee 1999). Although veteran state workers resisted vehemently (Lee 2007), their mobilization was thwarted due to their own fragmentation, and suppression from the enterprise and the government (Cai 2006). Ironically, when the public sector reform threatened the foundation of union organization and the livelihood of union members, enterprise unions in SOEs were ordered to assist the reform. Therefore, unions in the public sector are solely transmission belts. Their function is confined to conveying and implementing economic and ideological policies of the state, rather than representing and defending the working class with respect to economic restructuring.

Union Marginalization

With the rise of the private sector, the union marginalization approach argues that the reform has largely severed state control of the economy, allowing the private sector to create its own domain of dominance. Local governments were given full autonomy to promote economic growth by every means possible, including attracting foreign direct investment and encouraging local entrepreneurship. The literature on the local developmentalist state highlights the symbiotic relationship between local governments and enterprises (Walder 1995; Oi 1995; Wank 2001). These studies suggest that the state-capital collusion has led to a prioritization of economic growth over economic and workplace safety (Gallagher 2005; Lee 2006).

Although the central government managed to close legislative loopholes by enacting pro-labor legislation, its legislative thrust was largely blunted by law-evasion practices that are prevalent in the private sector; this has been acquiesced by governments at low levels. Enterprises, particularly those with no prior experience in unionization, tend to ignore the Trade Union Law. Even when a union is registered, it is prone to be marginalized. Gallagher (2005) explains that the prevalence of weak enterprise unions is due to managerial autonomy and a lack of will on the part of local governments to enforce the law:

> The state's withdrawal from its previous role as administrator of labor allocation and employment has granted enterprises a great degree of power in setting their labor practices. Attempt[s] to balance this withdrawal with greater attention to laws and regulations as a means of regulating managerial power have been mostly unsuccessful. Developmentalist local governments have neither the capacity nor the will to implement constraints on capital. The strengthening of worker organizations as a means to mitigate the unequal relationship between firms and individual workers has also not been achieved. (p. 96)

Dual Cooptation

The dual cooptation approach argues that unions are largely co-opted by the state and the capital in a marketizing economy. This approach argues that the state appears ambiguous with respect to the rising labor conflicts. The Leninist state has no intention of empowering independent unions to defend labor interests. Yet, the state-sponsored unionization attempts to create a firm-level labor institution under state control. The dilemma lies in how far the state allows the union reform to proceed, without allowing it to deviate from its primary goals of economic growth and social control.

Certain studies emphasize the role of official unions rather than enterprise unions (Chen 2003; Han 2010). In reality, official unions handle labor disputes on behalf of enterprise unions due to the latter's lack of authority and legitimacy in the eyes of employers and employees. Other research suggests that the capital is the co-determinant of union efficacy at the firm level, which significantly mitigates the impact of official unions on enterprise unions. Therefore, union registration and functioning are dependent on negotiation between the enterprise and official unions (He and Xie 2011; Liu, Li, and Kim 2013).

Recent research uses managerial Industrial Relation (IR) ideology to explain unionization outcomes. Liu and Li (2014) identified three types of managerial IR ideologies and their influence on the outcomes of unionization and union efficacy. They found that unionized enterprises tend to perceive state-led unionization as a political necessity or operational input. An enterprise allows a union to become established when it is imperative to cooperate with the ruling party's policy, or when the enterprise believes that enhancing employee involvement in corporate governance will be beneficial to the company's operation. However, Liu and Li (2014) primarily explore the conditions under which enterprises accept unionization, not the diversity of enterprise unions.

Gap in the Literature

After 30 years of economic reform, the state-centric approach must be re-examined because the state no longer controls the economy. In the field, I constantly encountered complaints from union cadres on the tremendous difficulty of persuading enterprises to comply with the Trade Union Law. Official unions provide guidance to enterprise unions about how to operate a union, how to prepare union paperwork, how to make regular reports of union activities to the supervising official union, and many other issues pertaining to union functioning. However, the extent to which the expectation of official unions can be met is dependent on the enterprise's continuing cooperation with them. Enterprises appoint a union chairperson, fund union operations, and provide personnel and logistical support to enterprise unions. While union registration has become a norm, union dues meet strong resistance. Therefore, the state influences primary unions in an indirect way, in the form of persuasion rather than coercion. While the managerial IR ideology, as applied in Liu and Li (2014), offers a fresh perspective on the role of management, this approach shows an inferential gap by assuming union efficacy is causally related to management's perception of the state-led unionization campaign. Enterprise unions are led and staffed by managerial personnel. All union chairpersons and committee members are managerial personnel or individuals with a managerial background. Union efficacy varies not with whether or not managerial personnel lead the union, but with how union leaders, who are usually managerial personnel, operate unions.

Therefore, this paper attempts to bridge the inferential gap between managerial IR ideology and union efficacy by introducing the agency of a union chairperson. The agent of enterprise union, the person who runs the union on a regular basis, is missing. The existing research displays a strong tendency to ignore the initiative of union leaders. The logic lies in the fact that because the enterprise has the right to appoint union leaders or to appoint union candidates to slate, the role of the union chairperson is relegated to the management's agent and is not counted as an explanatory variable. Based on my fieldwork, I argue that the initiative of a union chairperson is decisive in mobilizing resources and personnel in organizing union activities and collective actions. The position confers a sense of responsibility. Employees who are appointed to union leaderships are usually popular figures among their fellows. In SOEs, for example, union chairpersons are appointed from senior employees who are well respected. A similar pattern is observed in enterprise unions in the private sector, where employees who are seen as considerate, reliable, and responsible among colleagues are often appointed or elected as union chairpersons. Hence, my research focuses on union chairpersons as the primary explanation of union efficacy at the firm level.

Diversity at the Firm Level

The research identifies three types of enterprise unions—paperwork unions, managerial unions, and proto-economic unions, which demonstrate the variation of union efficacy. Paperwork unions represent a failed attempt of the state to install functioning unions in enterprises that are hostile to the state-led unionization. This happens when management prohibits unions from performimg substantial functions in order to avoid interference from managerial authority. Managerial unions are enterprise unions that play a limited role in performing traditional welfare functions, such as organizing recreational activities and distributing gifts during festivals. Their primary task is to assist the enterprise or other functional departments such as the human resource department to manage employees rather than to represent employees' interests solely. Nevertheless, this categorization does not lead to an underestimation of internal diversity. Managerial unions perform tasks that range from providing auxiliary services to an enterprise and organizing recreational activities for employees, to managing union welfare. An enterprise union is categorized as a proto-economic union when one of the following two conditions is satisfied: 1) the enterprise union is capable of engaging in wage bargaining with management, or 2) the enterprise union is directly involved in organizing or coordinating collective actions in order to maximize the economic interest of the employees. Proto-economic unions aggregate employees' interests in collective actions. Yet, the efficacy of proto-economic unions is determined by the extent to which union leaders can control collective action process.

Explaining Union Efficacy: The Enterprise's Perception and the Initiative of a Union Chairperson

Table 1 outlines the explanatory framework, in which the two explanatory variables are enterprises' perceptions of the state-sponsored unionization (IV-E) and the initiative of the union chairperson (IV-C). The value of IV-E varies between extraction and resource. The enterprise's perception of the state-sponsored unionization is a prerequisite for a union to function at the firm level. On the one hand, when enterprises primarily see the unionization campaign as a form of extraction, the operational space of the union will be largely diminished. In this case, management will likely appoint its loyal agent to serve concurrently as the union chairperson, which ensures total subordination of the union. On the other hand, perceiving unionization as a resource creates an incentive for enterprises to provide the union with operational space. Thereafter, the union chairperson can take the initiative. When the union chairperson acts as an agent of management, enterprise unions become managerial unions that assist enterprises with the management of employees. When the union chairperson acts as an agent of employees, the enterprise union evolves into a proto-economic union that can organize or coordinate collective action to promote the economic interests of the employees.

TABLE 1
Explanatory Framework of Union Efficacy: Enterprise's Perception and Union Agency

		Initiative of union chairperson	
		Act as the agent of management	**Act as the agent of employees**
Enterprise's perception	**Extraction**	Paperwork union (N = 3)	N/A
	Resource	Managerial union (N = 13)	Proto-economic union (N = 4)

What differentiates managerial unions from proto-economic unions is whether a union is capable of bargaining for substantial economic interest on behalf of employees, by organizing or coordinating collective actions. A union leader who has the ability to control collective action processes that bring parties with conflicting interests to the bargaining table is the litmus test of proto-economic unions. As a union leader's ability to control collective action increases, enterprise unions are more likely to wrestle concession from the capital and win recognition from the state. When a union leader is incapable of establishing a reputation among employees and is unable to control collective action processes, enterprise unions are unlikely to gain recognition from the capital and the state. Under these conditions, collective action carries a higher risk of failure. The following sections detail the explanatory process by explaining the role of the union chairperson and its impact on union efficacy.

Paperwork Unions

Paperwork unions represent the capital's determination to compromise state-sponsored unionization. Enterprises acquire union registration only to display ostensible compliance with the law and cease cooperation with the state after the registration. Paperwork unions develop under two conditions. First, the enterprise perceives state-sponsored unionization as an extraction and minimizes its harmful impact. Concomitantly, the union chairperson is appointed as a loyal agent by management to ensure the union's subordination to managerial authority. The enterprise union is thereafter prohibited from performing any function as stipulated in the law. LH's paperwork union is illustrative of how hostile enterprises that are determined to compromise state-sponsored unionization and use the appointment of union chairperson to effectively marginalize enterprise unions.

Unionization as State Extraction

LH, a Japanese-invested enterprise, acquired union registration in 2006 after being persuaded by the local official union. Since the registration, the firm has used every means possible to avoid union publicity. The most significant of these measures is the appointment of production and life team[1] members to double-sit on the union committee. The team leader is appointed as the union chairperson. The firm's hostility toward the union stems from a concern about union dues. Both the chairperson and the human resource manager regard union dues as a form of state extraction. For the enterprise, union dues are heavy burdens imposed by the state and a sign of government corruption. The human resource manager complained to me:

> Why [do] union dues have to be paid to the government (the official union)? The government is so corrupt. Nobody knows where the money would go. The wage bill of our firm is around 3 million. Paying 2% of 3 million as union dues means that we have to pay 160 thousand RMB per month and 720 thousand RMB per year. How could that be possible?

The union chairperson shared the same concern: "Fifty percent of … union due[s] will be refunded to the enterprise, but what about the other half? It will be confiscated by the state."

The Union Chairperson as an Agent of Management

The union chairperson then described the union as "being registered." Involuntary union registration is an implicit agreement between the firm and the official union. According to the chairperson, the firm is willing to cooperate with the government, as long as the cooperation does not affect the company's core interest. The enterprise is willing to have a paperwork union, as long as the unionization campaign imposes no real cost to the firm. The union chairperson then provided two reasons for why the firm maintains superficial cooperation with the official union. First, the firm does not want to leave an impression on employees that the company has a union. If employees know of the existence of the union, then they will use it as a vehicle to make more economic demands, such as bargaining for wage increases. For the union chairperson, these demands are excessive and will make management more difficult. Second, the union chairperson regards the union as a duplicated and redundant institution. For him, the union is similar to the production and life team of the enterprise. The two organizations have duplicate functions in terms of handling employees' complaints. Therefore, formalization of the union is not a worthy investment because the firm has an internal organization that specializes in these matters.

Managerial Unions

When firms see unionization as a resource rather than an extraction, enterprise unions are more likely to develop a formal structure and obtain limited recognition within the enterprises. Managerial unions operate in accordance with an enterprises' administration policy. The enterprise tends to see the managerial union as a supplementary department for addressing employees' trivial complaints, maintaining regular communication between management and employees, and handling logistical issues that fall outside of the duties of other departments.

Unionization as a Resource

Minimum compliance with the Trade Union Law provides firms with access to state resources. After balancing cost and benefit, these firms are willing to establish enterprise unions as an institution to receive state resources because of the corporatist nature of China's union system. The union system resembles an administrative pyramid of the government. By law, all unions must affiliate with the national peak labor organization—the All-China Federation of Trade Unions (ACFTU). The ACFTU legally represents all official unions and primary unions at the national level. Moving down the pyramid, every province has one provincial level union federation, as does each city. Union federations at the subdistrict level are further down the pyramid; these federations were created in 2006 as part of the downward penetration of the state. At a lower level, community-level union federations are established in areas with a high industry and working population. Enterprise unions lie at the lowest rung of this pyramid and constitute the overwhelming majority of primary unions.

Although the institutional design reflects the Leninist state's intention to control labor organizations, the market reform empowers actors in the private sector to effectively compromise this ostensible hierarchy. By law, enterprise unions must answer to the official unions that are one level above them. In reality, official unions have no legal or administrative authority to enforce the Trade Union Law on primary unions. Instead, official unions must rely on economic means to encourage enterprises to cooperate. One way for the official unions to induce cooperation is to distribute resources through the union system to unionized enterprises. The prerequisite for receiving resources is to establish an enterprise union. A number of enterprises acquire union registration in order to gain access to beneficial state resources.

SJB, an enterprise in the jewelry industry, established an enterprise union in 2010. "Amiable" is the word that was used by the union chairperson to describe their relation with the subdistrict union federation, which is supposed to be its supervising union. In her eyes, the official union is different from other state organs. The subdistrict union brings various benefits to the firm, including but not limited to recreational activities and vocational training funds. Vocational training on workplace safety and human resource management are in great demand by many enterprises that are eager to improve the quality of their workforce. Enterprises are reluctant to fund vocational training of employees out of their own pockets due to the relatively high cost of training sessions and the high turnover rate among employees. Nevertheless, with union registration, enterprises have access to a variety of these important resources at low or zero cost.

The state helps enterprises form a positive image of the unionization campaign. Sharing a similar view, the union's vice chairperson of FM, a Singaporean-invested enterprise in the electronic manufacturing industry, told me that the municipal union federation manages an amount of union funds that is worthy of a dozen million RMBs, specifically for vocational training. The official union makes annual plans for how to distribute these funds to enterprises. Last year, FM's union applied for subsidies from the official union in order to fund five employees to attend vocational training sessions on human resource management. When discussing union welfare, the vice chairperson beamed with delight: "The official union has done a good job. So far, what we need to do is to enjoy the welfare." The above cases suggest that enterprises are more likely to develop a positive attitude about unionization when they perceive the state campaign as a resource. Short of paying union dues, union registration channels a considerable amount of resources to enterprises at low or zero cost.

The Union Chairperson as an Agent of Management

In general, a managerial union has limited functions. The enterprise unions of SJB and FM primarily provide auxiliary services to enterprises or employees to smooth corporate governance. Managerial unions rarely fulfill their other obligations that are stipulated in the Trade Union Law. They have no bargaining power, nor are they eligible to participate in corporate decision-making, which could substantially affect employees' economic interests. In managerial unions, union chairpersons take the identity of a manager. For them, a union is necessary because it is either a legal requirement or a natural development of corporate structure. Nevertheless, a managerial union could become more effective at improving employees' welfare. A union chairperson's initiative largely determines the extent to which a managerial union can balance the interest among employees in managing and distributing union welfare.

In order to triangulate the explanatory framework in Table 1, a within-case analysis of managerial unions is employed to illustrate the role of a union chairperson in expanding the scope of union functions. Table 2 includes two more cases in order to strengthen the proposed causal inference of the impact of a union chairperson's initiative on union efficacy. A union chairperson, who is committed to union work, makes a union more effective at promoting union welfare. When a union chairperson has a weak initiative, the union tends to develop little specialization—performing only auxiliary functions.

TABLE 2
A Within-Case Analysis of Managerial Union

	Ownership	Enterprise's perception of the state-led unionization	Union chairperson's union ideology	Union chairperson's initiative	Relation with official union	Union efficacy (union welfare)
BM	State-holding	Guidance/ political necessity. Union dues are paid on a regular basis (the tradition of state-owned enterprises).	Union is to protect employees by supervising the enterprise to abide the law.	Weak initiative. Appointed by the enterprise.	Infrequent contact.	Weak/no specialization: provided logistical support to recreational activities organized by the enterprise.
FM	Foreign	Mixed: union welfare as a resource; union dues as state extraction. Management does not yet pay union dues.	Union is not much different from corporate management.	Strong initiative. Recommended by the ex-chairperson.	Infrequent contact.	Weak-medium/assist corporate governance—e.g., union suggestion box. Assertiveness: persuaded the boss to pay union dues a couple of times.
SJB	Domestic	Resource/guidance. Union dues are a burden.	Unionization is a "necessary trend." Union is a bridge that communicates employees to the company.	Weak initiative. Elected.	Infrequent but amiable.	Weak/no specialization: organized recreational activities.
YN	Foreign	Mixed: unionization as a means to improve employee loyalty; union dues as state extraction. Nevertheless, the enterprise pays union dues on a regular basis.	Union shall show solicitude to employees.	Full-time appointment. Strong initiative. Appointed by the senior manager.	Infrequent contact.	Strong/specialization: equalization of the distribution of union welfare. Institutional innovation: establishment of union groups in a chain store. Plan for the distribution and usage of union dues.

Case Study 1: Weak Initiative, Little Specialization

Previously a state-owned enterprise, BM is a publicly held company and a leading producer in the glass industry in China. Following the tradition of a state-owned enterprise, BM has had an enterprise union since the 1990s. The company's employment practice strictly abides by the Labor Contract Law and the Trade Union Law. The union elects union chairpersons and union committee members on a regular basis. In 2006, a new regulation of union organization was implemented, which banned the deputy director from sitting as the union chairperson. The firm immediately complied and removed the deputy director. BM's enterprise union, however, has not articulated a representative voice on behalf of employees. The union does not have a particular role to play in corporate governance, other than to organize some recreational activities. The union chairperson told me:

> The responsibility of union, as I understand it, is to protect employees. Enterprise unions shall stand behind employees when the enterprise violates the law. As an established firm with a law-abiding tradition, the enterprise has basically eliminated labor dispute from its root[s]. Now, the union barely needs to do anything. So long as there is no labor dispute, the union pretty much fulfills its mission.

This union chairperson's view reflects a narrowly defined role of a managerial union. Enterprise unions must supervise enterprises to ensure that they comply with the law, but they cannot go beyond the minimum legal requirements. When employees make economic demands on the enterprise above what the law stipulates, the union chairperson handles the claim in favor of the enterprise. The company had disputes about severance pay with senior employees, who worked for the firm for more than a decade. According to the Labor Contract Law, the compensation for one-year of employment is one month's average salary. Some senior employees demanded a larger amount of severance pay, given their long-term services. However, most of them lost their cases because the firm had provided compensation that was strictly in accordance with the law, and in some cases, higher than the amount prescribed by the law. During the entire dispute process, the union chairperson stood firmly in line with the enterprise's decision, rather than bargaining for a higher amount of severance pay on behalf of the employees.

Case Study 2: Strong Initiative, Increasing Specialization

When a union chairperson is willing to take the initiative to expand union welfare or innovate union organization, the managerial union becomes more effective in terms of developing a functional specialization, within the operational space allowed by the enterprise. YN is a catering company that owns more than 30 chain stores in its headquarters city. The enterprise union was registered in 2012, six years after the company was established. In the first two years, the enterprise union was only a paperwork union. The turning point came in 2014, when the general manager, one of the shareholders, decided to activate the union to substantiate the union functions. He appointed Ms. Tong (pseudonym) as the full-time chairperson. Ms. Tong told me that behind this decisive move was the general manager's concern about the high turnover rate of the employees, which resulted from, as he believes, the enterprise's excessive focus on profitability rather than cultivating employees' loyalty. The general manager expected the union and its full-time chairperson to specialize in handling employees' welfare, training, and career development.

Ms. Tong, previously an experienced operations manager at one of the chain stores, is committed to union work. A standard enterprise union has one chairperson, one vice chairperson, and five to six union committee members who are in charge of a particular aspect of union function. Soon after assuming the position, she began to innovate the union organization by creating a union team in each of the chain stores. The innovation was neither prescribed in the law nor initiated by the official union. Rather, it was an innovation conceived by the chairperson after consulting with the manager, in order to stimulate chain store employees' enthusiasm about participating in corporate social responsibility activities and to strengthen information exchange between the headquarters and its 38 chain stores.

Another significant move of Ms. Tong was to equalize the distribution of union welfare between headquarters employees and chain store employees. The distribution of union welfare has created much tension between some headquarters managers and Ms. Tong. The source of tension is the union dues paid by the enterprise. According to the Trade Union Law, a unionized enterprise pays an amount that is equivalent to 2% of its wage bill to the official union as union dues every month. In practice, in order to obtain support from an enterprise for unionization, the official unions must negotiate terms with employers to settle the acceptable amount of union dues paid by the enterprise. Despite employing more than 1,000 people, YN pays 2% of the wage bill of only 100 employees, who are mostly at the headquarters, as union dues. Subsequently, when Ms. Tong proposed to extend union welfare to chain store employees, her plan was immediately objected by certain headquarters managers who claimed that only those who have paid are entitled to the welfare. Despite the pressure from these managers, Ms. Tong insisted that the profit of the company is by contributed to by all employees rather than a small number of headquarters staff. Therefore, all employees are entitled to union welfare.

Proto-Economic Union

Proto-economic union represents the most effective enterprise union in the Leninist state. The term indicates an approximation of economic union/business union in the Anglo-American context, where trade unions fight for their members' economic interests by engaging in collective bargaining or economic strikes against employers. Table 3 compiles three cases of proto-economic unions to illustrate the causal connection between a union chairperson's ability to lead and control collective action processes and union efficacy in union-led collective actions.

TABLE 3
A Within-Case Analysis of Proto-Economic Union

	Ownership	**Enterprise's perception of the state-led unionization**	**Union chairperson's union ideology**	**Union chairperson's initiative (control of collective action process)**	**Relation with official union**	**Union efficacy (collective action)**
ZS	Foreign	Guidance/legal obligation. Union dues are paid on a regular basis before 2015.	The middleman: aggregation of employee opinion.	Effective. Elected. Consult, collect, and aggregate employees' opinions on wage increase.	Infrequent but informative.	Effective. Union has the priority to initiate wage bargaining, which is recognized by management.
ME	Previously domestic; now foreign	Legal obligation. Union dues have been paid on a regular basis since 2009.	Union is to protect employees.	Effective. Elected. Consult, collect and aggregate employees' demands and effectively control the collective action process.	Frequent. Maintained close contact with the official union throughout the collective action process.	Success. The enterprise made additional compensation to the employees, apart from the N+2 severance package.
SPG	Previously foreign; now joint venture.	Unknown. But the union was largely a managerial union before the strike. Union dues are not paid.	Union is a formal organization, which plays a key role in handling labor disputes.	Weak. Elected. Unable to aggregate employee's demands and unable to control the collective action process.	The official union does not recognize the elected vice chairperson.	Failure. The enterprise refused to offer severance pay to the employees. The union chairperson was subjected to administrative detention. Workers who participated the collective actions were dismissed.

Wage Negotiation

ZS is a small firm that specializes in original equipment manufacturer (OEM) watch manufacturing and employs 100 people. The firm is a subsidiary of the Montrichard Group, a global supplier for multinational corporations (MNCs), such as Disney and Avon. The enterprise union was established in 2011. The union chairperson, Mr. Xu (pseudonym), who was re-elected last year, is a veteran and a member of the Chinese Communist Party. When commenting on the primary achievement of the union, Mr. Xu emphasized that the union established a collective wage bargaining mechanism with management on behalf of assembly line workers. In 2014, the statutory minimum wage was increased to 1,800 RMB per month. Although at that time the base salary that ZS offered to its employees was significantly higher than minimum wage, Mr. Xu began to hear complaints from the employees. He then convened a meeting with assembly line workers and line leaders to obtain their opinions on the amount for a wage increase. Skilled workers demanded an increase of 350 RMB on top of their base salary, while general workers demanded 200 to 250 RMB. After an investigation into wage standards of the industry in the surrounding area, the union proposed an increase of 350 RMB for skilled workers and 250 RMB for general workers. The proposal was then written into a report, which was sent to the finance manager and then to the enterprise owner.

The enterprise owner soon sent a delegate to negotiate with the union, who then put forward the following two inquiries: Why is there a demand for a wage increase? And what are the reasons for the proposed amounts? Mr. Xu responded that the proposal accounts for three factors that affect employees' livelihood—a wage comparison with adjacent factories, the living expenses in the area, and the employees' demands. The union stressed the necessity of a living wage as the basis of the wage increase, suggesting that the delegate accept a wage increase at the level of a living wage. A living wage means that the employees would be paid sufficiently to support not only themselves but also their families. Management later conducted its own investigation and finally came to the same conclusion as the union.

Mr. Xu told me that a remarkable difference that the union brought is the transformation of individual claims into a collective claim. Prior to the establishment of the union, the company had no collective mechanism for wage negotiation. No one could effectively aggregate and convey employees' complaints and opinions to management. Employees had to resort to individual negotiations with a manager. When these individual demands were not met, workers engaged in stoppages or strikes to express their frustration. Now, the union has transformed the employees' unorganized and sporadic demands for higher wages into a coordinated collective action, which has significantly reduced incidents of stoppage and strikes.

Collective Action for Severance Pay

In the aftermath of 2008's financial crisis, there was a build-up of pressure on the export-driven manufacturing sector and a steady increase in the statutory minimum wage. The decline of export demand forced the state to revise its developmental strategy, which for a long time had relied on the labor-intensive and export-driven manufacturing industry. Local governments accelerated industrial upgrades by urging labor-intensive industries to relocate to less-developed regions or countries and revising foreign direct investment (FDI) policy in favor of capital-intensive and high-tech entrepreneurship. The years that followed saw a great decline in labor-intensive industries in the Pearl River Delta, the region where traditional manufacturing enterprises were once concentrated. Factories were closed. Production lines were removed. The demand for assembly line workforce decreased. The relocation of labor-intensive industry triggered a tide of labor disputes. Further aggravating the situation was the enactment of the 2008 Labor Contract Law, which significantly raised the cost of industrial upgrades for employers. This law stipulates that an enterprise is liable for severance pay when it lays off 20 employees, or 10% of its workforce when it employs fewer than 20 people. Employers then are liable to compensate one average month's salary to employees for each year of employment.

The role of enterprise unions in handling severance pay is rarely explored in existing research. This is due to the political taboo that any nonstate actor's participation and organization of collective actions will be either heavily restricted or punished by the Leninist state. Therefore, the existing research does not treat enterprise unions as relevant actors in analyzing the process and outcome of collective action. My fieldwork suggests that the enterprise union is capable of leading, organizing, and controlling the collective action process, provided that union chairpersons can control collective action processes to yield concession from the capital and win the recognition of the Leninist state.

Effective Union Leadership and the Success of Collective Action

ME, a subsidiary of TE Connectivity, specializes in designing and manufacturing telecommunication devices. The enterprise union was established in 1998. In 2009, the union held its first democratic election. Mr. Cao (pseudonym), an engineer from the Research and Development Department, was elected as the union chairperson. The following events suggest that the union chairperson played a crucial role in leading the union's collective action toward its success. Since 2010, ME began to lose its market share in China due to its inability to compete with native firms. In 2013, the company's annual sales suffered a constant decline, with an estimated deficit reaching eight million USD. In order to minimize the loss, the parent company announced its decision to dismantle the subsidiary's manufacturing and production lines. On September 10, ME announced its layoff contingency plan, which aimed to cut 500 out of 667 employees. The severance package proposed by the firm was N+2, meaning that ME would pay two extra months' salaries in addition to the amount calculated by the length of service.

Since the announcement of the plan, the union, led by Mr. Cao, undertook a series of measures to maximize the severance package. Initially, Mr. Cao collected information from other subsidiaries of the parent company in order to determine the maximum amount that the parent company was willing to offer for severance packages. He found out that the employees of one subsidiary, which will soon be relocated to another city, were offered 2N packages. Based on the information, the enterprise union demanded ME to offer a severance package of the same standard. But management rejected the union's first proposal immediately on the basis that the two subsidiaries are different. The union then proposed N+3 packages for employees who had worked for a minimum of five years and N+4 packages for those who had worked for over seven years. Again, the second proposal was rejected. The firm insisted on the original N+2 scheme. Knowing the management's firm stance, employees became restive and decided to launch a labor petition to the subdistrict government. This time, the employees demanded a negotiation with the representative of the parent company instead of the management of the subsidiary.

A labor petition carries a high risk in the Leninist state because the interests of the local state and enterprises are often intertwined, and collective actions often invite state suppression. Local government has a high stake in maintaining social stability and demonstrating a propensity to clamp down collective action of any sort. The successful leadership of Mr. Cao was the key to maintaining organizational discipline of the petition, which dissipates the government's anxiety and obtains its recognition of union legitimacy. On the morning of November 22, 200 employees marched toward the government building of the subdistrict after the enterprise refused to negotiate with them at the industrial park where the firm was located. The petition was well organized. The union chairperson demanded that all union committee members must use every means to maintain order among the employees with whom they are familiar. The marching crowd followed the union's order closely. After meeting with the government official, the union and the employees demanded that the representative of the parent company sit at the negotiation table. With the pressure from the government, the parent company finally decided to send a group of four delegates to negotiate with the union, the employees' representatives, and the local government.

On November 27, the two parties engaged in three rounds of intense bargaining. The primary contention was over the legality of the original N+2 scheme offered by ME. The delegates insisted that the N+2 package is strictly legal and that there would be no substantial concession that the enterprise could make. Mr. Cao refuted the claim. He told the delegates: "Abiding the law is the basic obligation of an enterprise. We, as senior employees who worked here for more than a decade, demand fair and reasonable compensation." Pressurized by the union's insistence and the strong claim made by the chairperson, the delegates finally came to an agreement with the union to increase the amount of compensation by adding an extra bonus that was calculated by the length of service. By the end of the bargaining, the parent company agreed to offer an extra 1 million RMBs to the employees as part of the severance package.

Weak Union Leadership and the Failure of Collective Action

When the union chairperson is neither capable of aggregating employees' interests nor able to control collective action processes, union mobilization may lead to its own defeat. The failed collective action of SPG highlights the necessity of union chairpersons to control collective action processes in a precarious social and political environment.

SPG is a process trade enterprise in the footwear sector. Previously an Australian-invested enterprise, its ownership was transferred to a Hong Kong investor in 2010. In 2013, after the second transfer of ownership, it became a joint-venture enterprise. Subsequently, the legal representatives of the firm were changed. The series of ownership transfer was triggered by a change in the local government's FDI policy, which now favors capital-intensive and high-tech entrepreneurship. In 2008, the provincial government issued a document that urged all process trade companies similar to SPG to undertake industrial upgrades. The third transfer of ownership was thus the compliance with the new FDI policy. On May 19, 2015, SPG formally announced to its employees that the enterprise had completed the ownership transfer. The firm promised to maintain the same level of salary and welfare of the employees, and, most important, the length of service would be continued after the transfer.

Most employees did not trust management. Their suspicion stemmed from the fact that the latest transfer had completely severed the tie between SPG and its parent company, an MNC listed in Australia, making SPG an independent firm with few fixed assets. Many of them regarded management's promise as a bounced check. The union's vice-chairperson, Mr. Niu (pseudonym), explained this anxiety to me: "The firm promises us that the salary and welfare will be maintained. But what do workers worry about? Now, the firm is just like a shell. It has no asset to guarantee the life of our employees." Mr. Niu's remarks reflected the worries of a large number of affected employees, who were soon motivated to demand severance pay. The severance pay later became not only a source of friction between employees and management but also a source of cleavage between the union and employees. SPG refused to offer severance pay to frustrated employees. Management insisted that the transfer does not affect the employees' financial interest, nor does the firm have any plans for relocation. Because the enterprise's plan of industrial upgrade completely abides by the law and the regulations of industrial upgrade, the employees' demands for severance pay were not reasonable. In response to the refusal, some workers petitioned the local government but were soon dispersed by the police. Others blocked the factory gate in order to pressure management into offering severance packages. Most others joined production stoppage.

The collective actions, however, were fragmented and lacked a centripetal force. Workers employed their own means of pressuring management with the hope of maximizing their respective gains, sometimes by creating chaos. In the process, no individual or organization, including the enterprise union, was able to effectively coordinate and aggregate the employees' demands and transform them into a collective voice. Two weeks after the outbreak of the stoppage, the enterprise union held a factory-wide democratic election, where Mr. Niu was elected the vice chairperson of the union. As a senior employee who had worked at SPG for over a decade, he had a positive reputation among his colleagues. Nevertheless, Mr. Niu could not control the collective action process. He and the union could do nothing to restrain the employees from demanding severance compensation. He told me:

> Severance compensation was not among the union's demand. We just talked about the guarantee. The company has no intention of relocation or layoff anyway. It is impossible to ask for a severance package, right? However, some employees are naïve. They are hoping to get the enterprise to make compensations before resigning the employment contract with them. We've told them again and again that compensation is unrealistic.

The union's suggestion, however, fell on deaf ears. In the first month of the stoppage, a large number of employees insisted on severance pay as a prerequisite for resuming assembly line production. Others engaged in more confrontational actions, including but not limited to occupying the office building of SPG. In spite of the enormous pressure that the workers tried to inflict on the enterprise, the enterprise did not yield or attempt to negotiate with the angry crowd. After the two-month protracted standoff between frustrated employees and management, the local police were involved to restore order in the factory; this ended the uncoordinated collective actions. In the end, management did not yield to the pressure of the employees. Many employees were dismissed during and after the stoppage. Many others resumed production. Some employees who were active participants of the occupations were put under administrative detention, and one of them was Mr. Niu. He was accused of instigating the workers to participate in illegal collective actions, which caused severe losses to the enterprise and undermined social stability.

The existing research has a strong tendency to disassociate collective action from enterprise unions created by the state-led unionization (Chen 2003; Chen 2010). This logic asserts that the Leninist state has crippled the ability of labor to organize and to mobilize autonomously by imposing unionization on enterprises. Enterprise unions are co-opted by management and are unable to demonstrate labor militancy. My fieldwork suggests that enterprise unions can be mobilized to organize or coordinate collective actions. There is a possibility that discontent employees use formal institutions to legitimize their confrontation with management. However, the success of collective action is highly dependent on the ability of a union chairperson to control collective action processes. Well-organized collective actions led by competent union leaders are not only capable of forcing concessions from the capital but also of gaining recognition from the state. By contrast, a poorly organized collective action with no center of leadership drains the patience of the Leninist state, leading to its own defeat.

Conclusion

This article explains union efficacy from an actor-centric perspective. Because state primary unions in the Leninist state are organized at the firm level, the structure-centric arguments about the state and the capital are not sufficient to explain the diversity of enterprise unions. This research contributes to an in-depth multiple-case study of enterprise unions, thus enabling us to evaluate the influence of firm-specific factors that shape the outcomes of state-sponsored unionization.

Endnote

[1] According to LH's union chairperson, the production and life team is an institution in many Japanese firms. The team that handles employees' complaints can be seen as part of the enterprise culture of Japanese enterprises.

Acknowledgments

This research is funded by the National University of Singapore.

References

Chen, F. (2003). "Between the State and Labour: The Conflict of Chinese Trade Unions' Double Identity in Market Reform." *The China Quarterly*, 176, 1006-1028.

Chen, F. (2010). "Trade Unions and the Quadripartite Interactions in Strike Settlement in China." *The China Quarterly*, 201, 104-124. doi:10.1017/S0305741009991093

Cai, Y. (2006). *State and Laid-Off Workers in Reform China: The Silence and Collective Action of the Retrenched.* London and New York: Routledge.

Feng, G. (2006). "The 'Institutional Weaknesses' of Labor Unions and Their Formative Contexts." *Society*, 26(3), 81-98.

Gallagher, M. E. (2005). *Contagious Capitalism: Globalization and the Politics of Labor in China.* Princeton, NJ: Princeton University Press.

Han, G. (2010). The Space for Institutional Innovations of CTU: Analysis of Integrated Right-Defense Model for Modern Workers in Yiwu City. www.sociology2010.cass.cn/upload/ 2012/08/d20120830132220761.pdf

He, B., and Xie, Y. (2011). "Wal-Mart's Trade Union in China." *Economic and Industrial Peace*, 33(3), 421-440.

Lee, C. K. (1999). "From Organized Dependence to Disorganized Despotism: Changing Labour Regimes in Chinese Factories." *China Quarterly*, 157, 44-71.

Linz, J. J. (2000). *Totalitarian and Authoritarian Regimes.* Boulder, CO: Lynne Rienner Publishers.

Lee, C. K. (2006). "Zhongguo gongren jieji de zhuanxing zhengzhi." In Y. Li and L. Sun (Eds.), *Social Stratification of Contemporary China: Theory and Evidence.* Beijing: Social Sciences Academic Press.

Lee, C. K. (2007). *Against the Law: Labor Protests in China's Rustbelt and Sunbelt.* Oakland, CA: University of California Press.

Liu, M. (2009). Chinese Employment Relations and Trade Unions in Transition. Doctor of Philosophy Dissertation, Cornell University.

Liu, M., Li, C., and Kim, S. (2011). "Chinese Trade Unions in Transition: A Three-Level Analysis." In P. Sheldon, S. Kim, Y. Li and M. Warner (Eds.), *China's Changing Workplace: Dynamism, Diversity and Disparity.* London and New York: Routledge.

Liu, M., and Li, C. (2014). "Environment Pressures, Managerial Industrial Relations Ideologies and Unionization in Chinese Enterprises." *British Journal of Industrial Relations*, 52(1), 82-111. doi:10.1111/j.1467-8543.2012.00908.x

Oi, J. C. (1995). "The Role of the Local State in China's Transitional Economy." *The China Quarterly*, 144, 1132-1149.

OEDC. (2012). *OEDC Employment Outlook 2012.* Paris: OEDC.

Walder, A. G. (1995). "Local Governments as Industrial Firms: An Organizational Analysis of China's Transitional Economy." *American Journal of Sociology*, 101(2), 263-301.

Wank, D. L. (2001). *Commodifying Communism: Business, Trust, and Politics in a Chinese City.* Cambridge, UK: Cambridge University Press.

III. Confronting Global Wage Stagnation

Older Workers and Wage Stagnation: Will 7.2 Million Older Workers Lower Your Wages?

Teresa Ghilarducci
Michael Papadopoulos
Siavash Radpour
The New School for Social Research

Introduction

Between 2014 and 2024, an extra 7.2 million older people are expected to enter the labor market because the numbers of older people will increase and their work effort is expected to rise. The first wave of 75 million baby boomers—born between 1946 and 1963—began to turn 65 in 2009, and an additional 10,000 baby boomers will turn 65 every day until about 2029 (Wyden and Ryan 2011). Boomers' labor force participation rates for the population over age 55 is predicted to rise from 45.9% to 50.7 % with the greatest rise for those between ages 65–69, from 31.7% to 39.5% 2024.

The unprecedented and large increase in the supply of older workers may continue to suppress the wages of the boomer cohort and may suppress or decrease wages of younger workers beyond what wage increases would have been expected. We use previous methods of estimating the effect of large labor supply cohorts on wages, which is estimating the elasticities of substitution of younger workers with respect to older workers. If older workers are substitutes for incumbent younger workers, their wages will fall. On the other hand, if older workers complement younger labor an increase in the supply of older labor will increase the demand for younger labor and increase their wages. Using a translog production function, we estimate how older workers may affect the wages of each other and younger workers as members of the baby boomer cohort age and continue to work more than in the past.

7.2 Million More Older Workers Will Work Between 2014–2024

The share of the labor force over age 55 is growing rapidly not only because of population aging but because the labor force participation rates of older workers are increasing. People in the labor force over age 55 grew by 10 million between 2004 and 2014, while people in the labor force under 55 fell by over 4.6 million. The number of people aged 55–74 is predicted to increase 14% between 2014 and 2024 (BLS 2015). Labor force participation rates for older workers have been rising since the 1980s. Between 1985 and 2013, the labor force participation rates for women ages 55 and older increased from 22% to over 35%; older men's participation rate rose from 41% to nearly 47% (BLS 2016).

Experts agree that the labor force is aging, but the predictions about how many older workers will enter or stay in the labor market vary. The variation is due to different predictions about future labor force participation rates of older workers. If the older-worker labor force participation rates do not change, the number of older workers will increase by 3.2 million between 2014 and 2024. The Bureau of Labor Statistics predicts a modest increase in the work effort of workers age 55–74 from 64% to 66% between 2014 and 2024, leading to a projection of 6.2 million more older workers in ten years (see appendix).

In contrast, Brookings economist Gary Burtless predicts that an additional 7.2 million people over age 65 will work. He predicts some combination of policy changes and changes in social norms and customs and preferences

Author's address: Teresa Ghilarducci, The New School for Social Research, 66 West 12th Street, New York, NY 10011 for work will increase older workers' labor supply beyond mere population increases. We compute the change in wages in both scenarios, although we believe Burtless has a better forecast than the BLS for a number of reasons.

Burtless' projections seem more likely than the BLS projection because the erosion of retirement income will likely induce more intensive and extensive labor supply among older workers). The labor supply of older workers is more sensitive to changes in the availability of unearned income than for younger workers; Cesarini et al. (2015) find a notable decrease in hours worked by lottery winners in Sweden, which persisted more than ten years—especially among older workers. California teachers aged 55 to 75 were found to be 2% to 3% less likely to work an additional year per $100,000 of accrued pension wealth. An unexpected increase in pension generosity was also associated with decreased work effort in this age group (Brown 2013).[1]

Employment Policy and the National Research Council

The National Research Council formed a study commission in 2012 motivated in part by a concern that the American economy would lose a vital supply of labor if everyone over 65 retired as boomers aged. The commission's report—"Aging and the Macroeconomy"—concluded that older workers were complements, not substitutes for younger workers. That conclusion was based on a 25-year old study by economists Phil Levine and Olivia Mitchell (Levine and Mitchell 1988). This paper updates their study and shows there was little support for the NRC's claim.

Levine and Mitchell use data from 1954 and 1984 to predict how increases in the share of older workers will affect the 2020 labor market in terms of the gender pay gap, job prospects for teenage workers, and wages of prime-age and young workers. Levine and Mitchell (1988) predicted the wages of various age and sex groups in the year 2020 based on projections about the age and sex composition of the workforce. They conclude older men's share in the labor force increase is associated with a fall in wages for most workers, especially females aged 16–19 and raise the wages of men aged 20–34. In contrast, they predict the increase in the share of older female workers would not decrease the wages of any sex/age group. They find support that older females (over age 55) significantly complement each other—an increase in the labor supply of older females raises the wages of older females.

Levine and Mitchell (1988) project the impact of an increase in the labor supply of baby boomers on wages for sex/age groups based on Bureau of Economic Analysis (BEA) and Social Security Administration (SSA) estimates for labor supply changes for those groups and their calculated elasticities of complementarity. Their estimates of labor supply elasticity may be marred by the BEA and SSA underestimating the actual growth labor supply from older workers. The BEA's growth rate for older workers' labor supply missed by over two-thirds, and the SSA's by almost one-half.

These prediction errors may explain why Levine and Mitchell (1988) overestimated wage growth among all groups except older workers. Increases in labor supply among younger groups were not met with a corresponding increase in demand, whereas older workers entered the labor market in larger numbers at a time when demand for them grew rapidly. Earnings for young workers decreased by 4.2% between 1985 and 2015 (Table 1), whereas Levine and Mitchell (1988) projected a 4.4–7.8% increase (Table 2). Moreover, mature workers' earnings increased 1.8%, compared to a projected 5.7–6.1% increase. Levine and Mitchell (1988) did not conclusively project older workers' earnings, with the SSA-based estimate projecting a 1.5% increase and the BEA-based estimate projecting a 24.3% increase. Older workers' earnings actually rose 8.3% in 30 years.

TABLE 1
BEA and SSA Projected Changes in Labor Supply by Age and Sex 1985–2020 vs. Actual 1985–2015

	BEA	SSA	Actual (BLS 2016)
Female Teen (16–19)	21.5%	19.5%	–33.4%
Female Young (20–34)	3.8%	4.1%	5.3%
Female Mature (35–54)	32.4%	47.3%	63.3%
Female Old (55+)	54.5%	92.0%	167.5%
Male Teen (16–19)	16.5%	18.4%	–43.4%
Male Young (20–34)	–9.9%	–4.7%	–1.0%
Male Mature (35–54)	23.5%	32.7%	45.4%
Male Old (55+)	46.1%	81.2%	109.6%

TABLE 2
Levine and Mitchell (LM) Real Earnings Growth Projections 1985–2020 vs. Actual 1985–2015 (Levine and Mitchell 1988)

	LM with BEA Population Projection	LM with SSA Population Projection	Actual (BLS 2016)
Teen (16–19)	5.6%	1.2%	0.5%
Young (20–34)	7.8%	4.4%	–4.2%
Mature (35–54)	6.1%	5.7%	1.8%
Older (55+)	24.3%	1.5%	8.3%
Female Workers	–10.3%	–7.9%	19.4%
Male Workers	12.2%	8.6%	–0.2%

The paper was occasionally referenced in the academic community, but it reached a high level of importance in 2013 because it was cited by an important National Research Council report supporting policies to increase older-worker labor market efforts. The National Research Council formed a study commission in 2012 motivated by the concern that the American economy would lose a vital and productive labor supply if everyone over 65 retired as boomers aged. Levine and Mitchell (1988) was the only paper footnoted to support the NRC's claim that older workers complemented younger workers. An updating of the study does not support the NRC's use of it as supporting evidence for the claim that older workers will not lower the wages of each other or younger workers.

Methods

This section updates Levine and Mitchell (1988) to challenge the NRC's conclusion that older workers working longer will not lower younger workers' wages. We follow the methodology used by Levine and Mitchell (1988) to reproduce their results using updated data.

> Hypothesis 1: If older workers are substitutes for incumbent younger workers, then the estimated cross-price elasticity will be negative—implying that younger workers' wages will fall as older workers enter the labor force.

Hypothesis 2: If older workers complement younger labor, then the estimated cross-price elasticity will be positive—implying that an increase in the supply of older labor will increase the demand for younger labor and increase their wages.

We use a translog production function in order to estimate the share of each factor of production in the total production cost, assuming that the output share of each factor will be equal to its cost. We run a system of seemingly related regressions using the factor share equations to estimate the coefficients, by imposing symmetry and homogeneity as constraints.

Instead of dividing labor by age into four categories (as in Levine and Mitchell 1988), we classify workers in three categories of young (34 years and younger), mature (35 to 55 years old), and old workers (55 to 74 years old). These groups are further divided by gender into six groups. The main reason for combining teen and young workers is the very small number of employed teenagers since the 1970s, mostly as a result of increasing years of schooling.

The quantity of each labor input, measured in hours worked, and estimations for hourly wages required for calculating the cost shares are obtained using CPS-ASEC (Annual Social and Economic Supplement) for years 1962 to 2014. Macroeconomic variables (GDP, wage share, and capital share) are obtained from BEA annual national accounts tables. We also use BEA's measure of "fixed assets" as a proxy of capital stock.

Similar to Levine and Mitchell (1988), the standard deviations and significance of estimated elasticities are calculated by using delta method.

The projected values for output and factor inputs in 2024 are from BLS labor force projections. The number of older workers, however, is adjusted according to the projections from Burtless. Average hours worked are held constant for the 2014–2024 period. The forecasted changes in wages for different age/sex groups are calculated directly by applying the estimated output share of each group to the projected GDP and labor quantities.

Results

Table 3 shows the elasticities of complementarity between workers in age/sex groups between 1962 and 2014. Because the changes are relative, a 1% increase in one category means a 1% decrease in the other one, and therefore elasticities are symmetric. Pairs including older male or female workers, which are of particular interest to the present study, are bolded. The diagonal depicts the effect of an increase in the share of an age/sex group on its own wages.

TABLE 3
Elasticities of Factor Complementarity 1962–2014

	FY	FP	FO	MY	MP	MO	K
FY	−5.74*** (1.32)						
FP	2.47*** (0.72)	−0.67 (0.66)					
FO	**0.03 (2.09)**	**2.50 (1.71)**	**−6.98 (7.23)**				
MY	2.95*** (0.85)	−3.78*** (0.47)	**−3.65*** (1.44)**	−1.41*** (0.57)			
MP	−1.01*** (0.40)	0.99*** (0.31)	**−2.45*** (0.79)**	1.07*** (0.26)	−1.19*** (0.17)		
MO	**−1.43** (0.85)**	**−0.96 (0.76)**	**5.79*** (2.87)**	**0.20 (0.57)**	**−0.20 (0.35)**	**−5.67*** (1.28)**	
K	0.39*** (0.17)	0.63*** (0.19)	2.03*** (0.52)	0.73*** (0.12)	0.76*** (0.09)	1.85*** (0.22)	−2.13*** (0.15)

Notes:
FY: Female Young (16–34) MY: Male Young (16–34) K: Capital
FP: Female Prime, Age (35–54) MP: Male Prime, Age (35–54) **$p < .05$
FO: Female Older (55–74) MO: Male Older (55–74) ***$p < .01$

Overall, out of 15 estimated elasticities between different labor groups, we find that four pairs are substitutes and five are complements. In contrast, Levine and Mitchell (1988) reported 13 significant coefficients out of 28 possible intergroup pairs, nine of which were substituting pairs and five of which were complementing pairs.

Wage Projections by Age and Sex

Table 4 displays two projections for wage growth for each age/sex group between 2014 and 2024. Both projections assume the same labor force participation rates for workers under age 55 throughout the period, as well as the same projected population by age and sex group (provided by the Census Bureau).

Using the BLS projections, we find that the real wages of young and old females, and prime-age and old males, will slightly increase over ten years. The wages of prime-age females will remain stagnant, and young male wages will decrease substantially. Burtless projects much larger increases in labor force participation among older workers than the BLS does. According to his projections, the wage loss and stagnation is more intense, and old males have a projected increase of 2.8% instead of the 8.0% increase using the BLS projections because they will compete with each other.

TABLE 4

Projected Percentage Change in Real Wages 2014–2024

	BLS	Burtless
Young Female	7.6%	4.3%
Prime-Age Female	0.5%	–0.5%
Old Female	3.7%	3.9%
Young Male	–5.3%	–6.7%
Prime-Age Male	8.4%	7.2%
Old Male	8.0%	2.8%

Notes: Young: 16–34, Prime-Age: 35–54, Old: 55+.

Hamermesh and Grant (1979) reviewed the literature on estimation methods used to address similar questions, including the methodology used in this paper. All of these methods use different production functions to derive quantity or cost functions, assuming the wages are determined in a competitive labor market with full employment, where the wages are equal to marginal cost of factors and their productivity. This is not an acceptable assumption when our aim is to estimate wages that may change according to the labor's bargaining power, which varies by unemployment level. Using a model that takes such frictions in the labor market into account can solve this issue, but it makes the estimation itself more difficult and even unfeasible. This is a useful area for future research.

Similar to Levine and Mitchell (1988), we interpret the results as if elasticity estimation does not imply any causation. A negative coefficient should be interpreted as a wage decrease associated with an increase in the relative supply of the other group. Because these elasticities are symmetric, practical interpretation requires determining which group is actually driving the change. We provide justification in our Introduction for determining that older workers will predominantly drive the changes in younger workers' wages and not vice versa.

Another underlying assumption is that labor supply changes are exogenous, and wages react to labor supply without any feedback effects. This is an acceptable assumption when we use time-series data rather than cross-sectional data. Cross-section data is prone to change in labor supply due to migration across geographical boundaries due to wage differences. On the other hand, using time-series data requires us to assume that elasticities of complementarity are fixed from 1962 to 2024.

But, as a result of a change in technology and structure of industries in the period and changes in the nature of jobs that can be fulfilled by old and young men and women, this assumption is not likely to be true. For instance, employment in health care and social assistance, where older women are more likely to find work, is predicted to grow 1.9% a year from 2014 to 2024 (BLS 2015). Assuming these changes are exogenous may lead to weak and invalid estimations. The labor productivity of each sex and age group has changed during the time period, and we can only take the change into account using a dynamic model.

This problem is also related to factor classification. We divide the labor force by gender and age because we believe there is a meaningful difference between the groups, and not within a specific group, during the period of study. However, educational attainment for all groups has increased drastically. Older workers and women are more educated in 2020 than they were in the 1960s, especially relative to prime-age workers. It's fair to assume that the "new" older workers are now able to compete with educated prime-age workers for jobs, whereas the previous generations didn't have the necessary education and skill to do so. Dividing the factors into smaller groups based on education would make more sense and could also limit the amount of change in relative productivity of factors (see Grant 1979 for an example). If we were to break down each group by education, such problems would disappear. However, the small sample size does not allow a doubling of the number of factors. It is possible, however, to expand the study using the same sample by including measures of human capital of each age and gender group instead of the simple supply of labor. This goes beyond the scope of this study.

As a result of problems mentioned above, mainly the changes in factor productivity over time, in addition to our small sample size, elasticities of complementary are sensitive to the sample choice. Running the regression over smaller subsamples shows that the estimations, especially those for the earlier years, can be volatile, though it is hard to know if the problems are only caused by the even smaller size of the subsample.

Discussion

Burtless and Aaron (2013) describe the two ways of increasing labor supply—raising wages and improving working conditions, or reducing nonlabor income to induce more work—by categorizing pro-work policies for older people as either "mugging" and "bribing" policies. Mugging policies aim to reduce nonlabor income, including reducing Social Security benefits by raising the Social Security full retirement age and Medicare eligibility to age 70. Bribing policies range from the SSA emphasizing the rewards from delaying collecting benefits to age 70 to raising the after-tax wage by reducing taxes for older workers.

The increase in expected retirement age is consistent with the prediction that pension insecurity will increase the work effort of older Americans. The expected retirement age has been rising since 1991 when 11% of workers polled expected to retire after age 65. In 2015, the highest percentage of workers ever polled—36%—said they will retire after age 65 since the EBRI began the Retirement Confidence Survey in 1980 (Helman et al. 2015).

The negative wage effects of older workers entering the labor market are confirmed by the evidence that immigrants can lower the wages for some groups they directly compete with in the short term. Most economists agree that immigration has distributional effects (Chassamboulli and Palivos 2014). Immigrants lower wages and working conditions for native workers who have jobs that immigrants can easily fill such as those that require less language fluency or training in the short run (Dustmann et al. 2013; Liu 2013; Smith 2012).

Wages are stagnating across all age groups, and the increase in the labor supply of older workers may be a reason. In the next seven years, 7.2 million older workers, an increase in the overall labor force of almost 5%, are predicted to migrate from a planned or hoped-for retirement into the U.S. labor market. The magnitude of this surge of "domestic migrants" compares to that of other periods in American history when large increases in the labor supply reduced workers' bargaining power to achieve wage growth or improvements in working conditions.

FIGURE 1
Older Workers' Labor Supply Is Similar to Historic Episodes of Labor Shocks

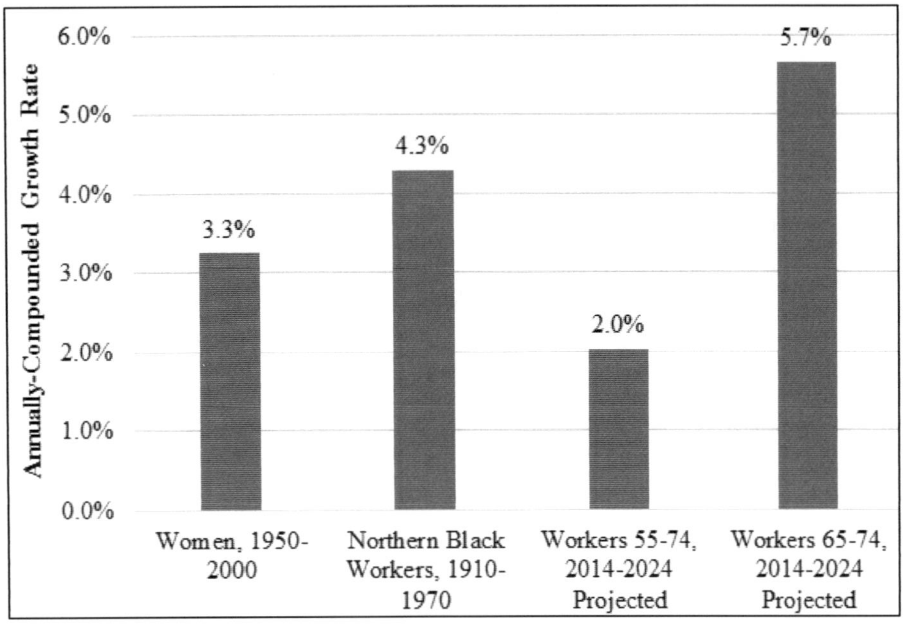

Sources: Census Bureau (1910, 1950, 1970, 2000).

Using the Burtless predictions, we expect the labor supply of people aged 55 to 74 to grow 2.0% each year and that of workers aged 65 to 74 to grow 5.8% each year between 2014 and 2024. This compares to the historic increases in the supply of black and women workers. The number of black workers in the North grew by an average of 4.3% each year between 1910 and 1970. The pre-WWII wave of the Great Migration from the American South to the North was found to have lowered northern black wages by about 3% (Boustan 2009). Furthermore, the labor supply of women grew 3.3% each year between 1950 and 2000. Pryor and Schaeffer (2000) found that increasing supply of educated white women workers after 1950 lowered the wages of less educated men. Many researchers have documented negative cohort effects, especially regarding the baby boom generation, to conclude boomer wages would have been higher if the cohort had been smaller (McMillan and Baesel 1990; Sapozhnikov and Treist 2007).

Evidence for less bargaining power of older workers is already showing up in wages, long-term unemployed, and decrease in job quality. Older workers make up a larger share of the long-term unemployed than ever before, workers over 55 earn substantially less than people in their late forties (Guvenen et al. 2015), and quality of jobs older people hold has stopped improving. The share of older workers who say they have very physically demanding jobs is increasing and the share of jobs reported as easy is falling. The incidence of requirements for stooping, bending, and using keen eyesight and intense concentration is increasing (Bonen 2013).

The true effect of migrants is measured by what conditions would be like if they had not come. Danish economists directly tested the wage effects of immigrants on native worker wages and working conditions. Denmark is far away, but Denmark is a good source for US labor market hypotheses because the Danish data match workers to employers, which allows direct observation of firms hiring immigrant workers. Incumbent workers are paid lower wages than they would otherwise in firms that hire immigrant workers (Malchow-Moller et al. 2012).

An Immigrant Bonus?

Alicia Munnell and April Yanyuan Wu (2012) argue that the theory new workers will lower wages, hours, and working conditions is a "lump of labor fallacy." Indeed the argument dates to 1851 when groups advocated for shorter weekdays—essentially Sunday off—by arguing that fewer hours would create more jobs. In modern terms, it is a fallacy to believe if older workers retired more jobs would be available for younger workers. They argue in particular that "there's no evidence to support that increased employment by older people is going to hurt younger people in any way. … It's not going to reduce their wages, it's not going to reduce their hours, it's not going to do anything bad to them."

Munnell and Wu can draw support from David Card's study of the immediate effects of the Mariel boatlift on incumbent Cuban-American workers. In 1980 over 125,000 Cuban refugees, former prisoners, and mental health patients increased the supply of mostly unskilled labor in the Miami labor market by 7% in just two months, and Card (1990) argues this had no effect on wages. However, Borjas (2015) found that the "absolute wage of high school dropouts in Miami dropped dramatically implying an elasticity of wages with respect to the number of workers between –0.5 and –1.5."

Regardless, a major contrast between immigrants entering a labor market and older "domestic migrants" is the state of the labor market they are entering. An influx of immigrants is associated with buoyant labor markets because migrants tend to come when conditions are good. We argue that older workers will not necessarily be entering the labor market because it is good, rather they will be coerced into the labor market, regardless of its state, due to their inability to afford retirement. Doubtless, the increase in labor supply will lead to economic growth and an absolute increase in the number of jobs available. The paper argues that without the increase in labor supply, wages, hours, and working conditions would be improved for all workers.

We are not committing a lump of labor fallacy. An increase in the number of older workers might increase the number of jobs available by bringing down wages. We argue that solving the retirement crisis will mitigate a 7.2 million worker labor supply shock.

Conclusion

The increase of 7.2 million workers over a ten-year period will affect labor markets, but will the effect be large enough to make a difference in the wages of younger workers? The answer is most certainly yes.

The macroeconomic case for policies to increase the labor supply of older workers may have ignored the negative microeconomic impact on the labor market for all workers. The NRC's 2012 report concluded that increasing the labor supply of older workers would not decrease the wages of younger workers. The report cited a 1988 paper to make the claim. This study updates Levine and Mitchell's (1988) mbitious study of the elasticities of complementarity with respect to the supply of workers by age and sex categories to confirm some of their findings and find new ones. In many cases, older workers significantly substitute for younger workers and with each other, resulting in lower wage growth for younger workers than would have otherwise occurred. Increasing the security and adequacy of American workers' pensions will reduce the labor supply of older workers and help make the wages of prime-age workers grow.

Endnote

[1] Increases in the net-of-tax wage through the Earned Income Tax Credit increases intensive work effort of Americans and British workers particularly among those over age 65 (Blundell et al. 2013). And older workers are more sensitive to temporary net-of-tax wage changes (so-called Frisch elasticities) than are younger workers (Reichling and Whalen 2012).

References

Blundell, R., Bozio, A., and Laroque, G. (2013). Extensive and Intensive Margins of Labour Supply: Work and Working Hours in the US, the UK and France. *Fiscal Studies*, 34(1), 1-29.

Borjas, G. J. (2015). *The Wage Impact of the Marielitos: A Reappraisal* (No. w21588). National Bureau of Economic Research.

Bonen, A. (2013). Older Workers and Employers' Demands. *Policy Note. The New School: Schwartz Center for Economic Policy Analysis*, 2007-20.

Boustan, L. P. (2009). Competition in the Promised Land: Black Migration and Racial Wage Convergence in the North, 1940–1970. *The Journal of Economic History*, 69(03), 755-782.

Brown, K. M. (2013). The Link Between Pensions and Retirement Timing: Lessons from California Teachers. *Journal of Public Economics*, 98, 1-14.

Bureau of Labor Statistics. (2015). Economic News Release: Employment Projections: 2014-24.

Bureau of Labor Statistics. (2016). Occupational Employment Statistics. Accessed March 16, 2016. www.bls.gov/oes/

Burtless, G., and Aaron, H. (Eds.). (2013). *Closing the Deficit: How Much Can Later Retirement Help?* Brookings Institution Press.

Card, D. (1990). The Impact of the Mariel Boatlift on the Miami Labor Market. *Industrial and Labor Relations Review*, 43(2), 245-257.

Census Bureau. *1910 Census*. U.S. Census Bureau. 1910. www.census.gov/prod/www/ decennial.html#y1910

Census Bureau. *1950 Census*. U.S. Census Bureau. 1950. www.census.gov/prod/www/ decennial.html#y1950

Census Bureau. *1970 Census*. U.S. Census Bureau. 1970. www.census.gov/prod/www/ decennial.html#y1970

Census Bureau. *2000 Census*. U.S. Census Bureau. 2000. www.census.gov/prod/www/decennial.html#y2000

Cesarini, D., Lindqvist, E., Notowidigdo, M. J., and Östling, R. (2015). *The Effect of Wealth on Individual and Household Labor Supply: Evidence from Swedish Lotteries*. National Bureau of Economic Research Working Paper.

Chassamboulli, A., and Palivos, T. (2014). A Search–Equilibrium Approach to the Effects of Immigration on Labor Market Outcomes. *International Economic Review*, 55(1), 111-129.

Dustmann, C., Frattini, T., and Preston, I. P. (2013). The Effect of Immigration Along the Distribution of Wages. *The Review of Economic Studies*, 80(1), 145-173.

Grant, J. (1979). Labor Substitution in US Manufacturing. Unpublished Ph.D. dissertation, Michigan State University.

Guvenen, F., Karahan, F., Ozkan, S., and Song, J. (2015). *What Do Data on Millions of US Workers Reveal About Life-Cycle Earnings Risk?* (No. w20913). National Bureau of Economic Research.

Hamermesh, D. S., and Grant, J. (1979). Econometric Studies of Labor–Labor Substitution and Their Implications for Policy. *Journal of Human Resources*, 518-542.

Helman, R., Copeland, C., and VanDerhei, J. (2015). The 2015 Retirement Confidence Survey: Having a Retirement Savings Plan a key factor in Americans' Retirement Confidence. *EBRI Issue Brief* (413).

Levine, P. B., and O. S. Mitchell (1988). The Baby Boom's Legacy: Relative Wages in the Twenty-First Century. *American Economic Review*, 72, 66–69.

Liu, C. Y. (2013). Latino Immigration and the Low‐Skill Urban Labor Market: The Case of Atlanta. *Social Science Quarterly*, 94(1), 131-157.

Malcho–Møller, N., Munch, J. R., and Skaksen, J. R. (2012). Do Immigrants Affect Firm‐specific Wages? *The Scandinavian Journal of Economics*, 114(4), 1267-1295.

McMillan, H. M., and Baesel, J. B. (1990). The Macroeconomic Impact of the Baby Boom Generation. *Journal of Macroeconomics*, 12(2), 167-195.

Munnell, A. H., and Wu, A. Y. (2012). *Are Aging Baby Boomers Squeezing Young Workers Out of Jobs?* Center for Retirement Research at Boston College, Working Paper 1218.

National Research Council. (2012). *Aging and the Macroeconomy. Long-Term Implications of an Older Population*. Committee on the Long-Run Macroeconomic Effects of the Aging U.S. Population. Board on Mathematical Sciences and their Applications, Division on Engineering and Physical Sciences, and Committee on Population, Division of Behavioral and Social Sciences and Education. Washington, D.C.: The National Academies Press.

Pryor, F. L., and Schaffer, D. L. (2000). *Who's Not Working and Why: Employment, Cognitive Skills, Wages, and the Changing US Labor Market*. Cambridge, Ma: Cambridge University Press.

Reichling, F., and Whalen, C. (2012). *Review of Estimates of the Frisch Elasticity of Labor Supply*. Washington, DC: Congressional Budget Office.

Sapozhnikov, M., and Triest, R.K. (2007). *Population Aging, Labor Demand, and the Structure of Wages.* Federal Reserve Bank of Boston Working Paper 07-8, Boston, MA.

Smith, C. L. (2012). The impact of Low-Skilled Immigration on the Youth Labor Market. *Journal of Labor Economics*, 30(1), 55-89.

Wyden, R., and Ryan, P. (2011). *Guaranteed Choices to Strengthen Medicare and Health Security for All: Bipartisan Options for the Future.* US House, Budget Committee.

Appendix A

Three scenarios regarding the potential increase in the labor supply of older workers are estimated below.

TABLE A.1
Increase in Labor Supply by Age If There Are No Changes in Labor Force Participation Rates in Any Age Category

	Projected LFP, 2024		**Labor Force (1000s)**	
Age	**2024**	**+/–, 2014**	**2024**	**+/–, 2014**
55–59	71.5%	0.0%	14,516.5	
60–61	63.4%	0.0%	5,453.9	928.86
62–64	50.2%	0.0%	6,460.4	
65–69	31.6%	0.0%	6,280.1	2,539.42
70–74	18.9%	0.0%	3,104.4	
	45.9%	0.0%	35,815.3	3,468.3

TABLE A.2
Increase in Labor Supply Based on Changes in Labor Force Participation Predicted by the Bureau of Labor Statistics

	Projected LFP, 2024		**Labor Force (1000s)**	
Age	**2024**	**+/–, 2014**	**2024**	**+/–, 2014**
55–59	74.2%	2.7%	15,069.0	
60–61	67.2%	3.7%	5,774.1	2,190.47
62–64	53.2%	3.0%	6,849.4	
65–69	36.2%	4.6%	7,203.0	4,036.56
70–74	22.4%	3.5%	3,678.5	
	49.4%	3.5%	38,574.0	6,227.0

TABLE A.3
Increase in Labor Supply Based on Increased
Labor Force Persistence Predicted by Burtless (2010)

	Projected LFP, 2024		Labor Force (1000s)	
Age	**2024**	**+/–, 2014**	**2024**	**+/–, 2014**
55–59	73.5%	2.0%	14,919.7	2,184.16
60–61	65.0%	1.6%	5,590.9	
62–64	55.8%	5.6%	7,175.6	
65–69	39.3%	7.8%	7,824.2	5,026.47
70–74	24.7%	5.7%	4,047.3	
	50.7%	4.8%	39,557.6	7,210.6

IV. LERA/ISA Joint Session—The Evolving Health Care Landscape: How Employees, Organizations, and Institutions Are Adapting and Innovating—LERA Research Volume 2016

Health Care Providers and Patients in Sync: Antecedents for Optimizing Provider and Patient Outcomes

Deirdre McCaughey
University of Alabama at Birmingham
Department of Health Services Administration

Gwen McGhan
University of Alabama at Birmingham
School of Nursing

Background

November 1999 saw the first definitive movement toward pursuing better health care outcomes with the release of the Institute of Medicine's (IOM) seminal report on health care quality, *To Err Is Human* (IOM 1999). The report focused attention on the estimated number of annual deaths attributable to preventable lapses in safety and served as a catalyst in driving patient safety to the forefront of industry focus. Three salient facts emerged from the IOM (1999) report. First, safety lapses ranked as the 8th leading cause of death in the United States. Second, the estimated cost of these lapses was approximately $29 billion annually. Third, most of these preventable errors derived from systemic errors rather than being the "fault" of a care provider (Gandhi 2015). Since the release of the report, a robust and continually evolving patient safety movement has witnessed the emergence of think tanks and institutions devoted to improving processes and systems in pursuit of error-free patient care.

Simultaneously, the rate of employee injury and illness in the health care sector in the United States rose substantially. Over the past decade, the reported number of work derived illnesses and injuries in health care have been among the highest rates of all industries in the United States. Shockingly, the health care sector accounts for the greatest percentage (20.7%) of private industry nonfatal occupational injuries among all industry sectors (Gomaa et al. 2015). For example, in 2013, nurses and nursing assistants in state/government institutions had occupational injury rates of 13.6 cases and 19.2 cases per 100 workers, respectively, while the national average across all industries was 3.7 per 100 workers (Bureau of Labor Statistics 2013). Given these numbers, it is not surprising a recent United States Department of Labor Secretary identified the health care industry as a major source of all U.S. workplace injuries: "We remain concerned that more workers are injured in the health care and social assistance industry sector than in any other, including construction and manufacturing" (Occupational Safety and Health Administration 2011).

Working within the health care industry has been shown to be inherently dangerous. Despite the staggering rate of injury and illness and its enormous cost to the health care industry, no report about the occupational wellness of the health care workforce has catalyzed an industry response as *To Err Is Human* (1999) did for patient safety. As

Author's address: Deirdre McCaughey, Department of Health Services Administration, University of Alabama at Birmingham, 580E SHP, 1720 2nd Avenue South, Birmingham, AL 35294-0113 such, the purpose of this paper is to propose an overarching model to integrate research on patient safety and employee safety so that both may be addressed simultaneously.

The Patient Safety Movement

Following the release of the IOM (1999) report, health care organizations were encouraged to develop an organizational environment to support the definition and creation of a culture in which patient safety was a critical organizational priority (Sammer, Lykens, Singh, Mains, and Lackan 2010). This safety movement was further bolstered by government agencies and health care organizations demanding patient safety become a key indicator of health care quality. Leading this initiative, the Joint Commission established the National Patient Safety Goals and promoted systematic tools, such as Failure Mode and Effects Analysis, to foster proactive examinations of systems and processes that contribute to safety failure (Joint Commission Resources 2010). The Agency for Health care Research and Quality (AHRQ) adopted the following definition of patient safety culture, which remains widely used throughout the industry (Sammer et al. 2010):

> The safety culture of an organization is the product of individual and group values, attitudes, perceptions, competencies, and patterns of behavior that determine the commitment to, and the style and proficiency of, an organization's health and safety management. (Health and Safety Commission Advisory Committee on the Safety of Nuclear Installations 1993)

Using this foundational definition, AHRQ developed and instituted the patient safety measurement tool "Hospital Survey on Patient Safety Culture" in 2004. Since its initial launch, AHRQ has continuously adapted the tool in an attempt to measure the "safety culture" within organizations including hospitals, medical offices, ambulatory surgery centers, and nursing homes. The patient safety culture tool measures a variety of sub-areas and serves to help health care organizations better understand their employees' beliefs about the culture of safety and how it interacts with patient care. While there are a number of safety culture tools and components, Sammer and colleagues' (2010) literature review categorizes the general properties that contribute to creating a robust patient safety culture. The categories include leadership, teamwork, evidence-based care practices, communication (e.g., speaking up), learning (the organization's learning capacity), just (errors are recognized as system rather than individual failures), and patient-centered care. These factors are critical to organizations (or units) creating an optimal patient safety culture in which preventable adverse patient safety events are reduced and outcomes are optimized (Sammer et al. 2010).

Employee Safety Climate

Safety in the realm of patient outcomes and care is delineated by the term "safety culture." Safety culture (also referred to in the organizational literature as safety climate; see paragraph below) is meant to describe the values, norms, and assumptions held by employees regarding the safety in an organization (Reiman and Rollenhagen 2014; Sammer et al. 2010). These deeper values are often reflected in and measured by shared perceptions of existing safety policies, procedures, and practices (Flin 2007). That is, safety climate (or culture) is a snapshot in time of employee beliefs about safety, linked to specific and identifiable policies and practices (Flin 2007). Thus, in this paper, we use the term "safety climate" when we refer to facets of both patient safety and employee safety.

Specific to employees' well-being, a climate of safety exists within organizations and reflects employee perceptions of the organizational climate and leadership, as well as experiential factors (e.g. being injured at work). Employee perceptions foster individual opinions of the workplace as potentially harmful or safe for the individual (Carr, Schmidt, Ford, and DeShon 2003; Parker et al. 2003). Referred to as "employee safety climate" in the occupational health and safety literature, it reflects a broad concept encompassing employee perceptions of safety practices, safety knowledge, organizational safety policies, and safety training among other factors (McCaughey DelliFraine, McGhan, and Bruning 2013; Neal, Griffin, and Hart 2000). In keeping with the established norms of the extant occupational health and safety literature, in this paper we use the term employee safety climate when we refer to facets of employee health and safety.

From a holistic perspective, health psychologists argue that employee safety climate is a critical component of "healthy" work environments (Danna and Griffin 1999). Workplaces characterized by high injury risk, violence, high strain, and chronic stress have been linked to adverse health outcomes for the employees such as burnout, high injury rates, and depression (Danna and Griffin 1999; Clarke 2006). Additionally, a workplace-derived injury or illness can shape employee climate perceptions and influence subsequent workplace performance/behavior (Colley, Lincolne, and Neal 2013; McCaughey et al. 2014).

Given the high injury/illness rates of health care workers in the U.S. and the negative effect a poor employee safety climate has on employee and organizational outcomes, it seems intuitive to bring it to the forefront of organizational priorities in health care. Equally deserving focus is a research agenda examining and integrating the factors of patient safety with facets of employee safety thus providing organizations with a synergistic opportunity to optimize safety outcomes for all.

Flin's (2007) review of the health care safety literature identified inconsistencies across safety climate definitions, the interchangeable use of the terms safety climate and safety culture, and the lack of specific theoretical work to delineate antecedents of employee and patient safety. Also identified in the review was a lack of health care specific studies examining patient safety and employee safety climate concurrently and its relationship with negative outcomes for employees and patients. In highlighting industry best safety practices and evaluating them with regard to health care studies, Flin (2007) proposed the Model of Safety Climate and Injury Outcomes. This model attempts to address the limitations noted above while accounting for the duality of employee/patient safety outcomes with Flin's use of the term "safety climate" to encompass the dual ethos of employee safety climate and patient safety climate. The existing safety climate may contribute to motivation (e.g., expectations for outcomes of particular behavior) and unsafe behaviors (e.g., rule breaking and risk taking), potentially resulting in errors leading to patient and/or worker injury (Flin 2007). The frequency and degree of employee and patient injuries necessitate the need for this type of model, allowing for the examination of adverse event pathways and the determination if processes contributing to errors and injury are the same for employees and patients.

Integrating Patient and Employee Safety Climate

Flin's (2007) review of the safety climate literature is one of the first to propose similar antecedent pathways that lead to both patient adverse events (e.g., "never events," which are serious, but largely preventable, clinical events (National Quality Forum 2011) and employee adverse events (e.g., back injury). Here, existing information on safety climate is adapted for the health care environment. The result is the establishment of an antecedent safety climate that is defined by employee perceptions of the prioritization of safety at the organizational level (senior management) and the department/unit level (supervisor). These beliefs may then serve as the primary antecedent for employee safety motivation and subsequent behaviors, which in turn may lead to errors that result in adverse safety events for patients and employees.

Employees in health care organizations are often faced with difficult choices; when pressed for time should one follow a more time-consuming safety protocol or engage in an action that may be more expeditious but puts the patient and care provider at increased risk for an adverse event or injury? Low commitment to safety procedures results in taking unsafe actions that may be the starting point for an adverse event (Flin 2007). Integrating the model to examine patient and care provider safety provides a framework for health care organizations to influence individual care provider behavior while targeting safety improvement for both members of the safety dyad.

Although Flin's (2007) model highlights the potential for integration, research examining safety climate has been slow to do so. There is, however, a beginning movement by prominent institutions to recognize the synergistic relationship between the outcomes for employees and the patients for whom they provide care. The Joint Commission's monograph, *Improving Patient and Worker Safety* (2012), identifies synergies for addressing safety climate for patients and care providers. Specific to the health care industry, the Occupational Safety and Health Administration (OSHA) has identified organizational safety climate as being the key foundation for linking patient and employee safety programs and outcomes (OSHA, n/d). Finally, the National Patient Safety Foundation has specifically linked the climate and environment of health care organizations to patient and care provider outcomes (Gandhi 2015). These industry leaders may provide the needed catalyst for a call to action regarding care provider safety as *To Err Is Human* (1999) did for patient safety.

Next Steps: Integrating Safety Outcomes

An agent to assist this catalyst is a guiding framework for integrating and identifying mechanisms that minimize adverse safety events for patients and care providers. While instrumental to understanding the individual pathways for employees and patients, Flin's (2007) model does not capture the multiplicity of workplace factors in adverse safety events. Such factors include staffing levels, process training, safety training, team cohesiveness, coworker support, and the environment for safety actions (such as stopping processes and error reporting). To address the complexity of the environment, a framework accounting for organizational influences on employee actions is critical to a research agenda linking patient and care provider safety.

The National Institute for Occupational Safety and Health (NIOSH) framework was developed to capture workforce factors and to identify how organizational practices influence work-life quality and the subsequent effect on employee on-the-job safety and health. NIOSH's framework guides occupational safety and health research by examining the factors contributing to employee illness and injury (Sauter et al. 2002). In this framework, the pathways to employee adverse safety events are posited to be similar to patient adverse safety event pathways. Here, occupational illness/injury occurs as a result of exposure to "psychological stressors" and "physical hazards." Psychological stressors include common work factors such as high demand/low control, role conflict, and fatigue. Physical hazards include ergonomics, exposure to pathogens, over exertion, and workplace violence. Physical and psychological hazards exist in organizations that potentially foster adverse patient safety events in the same manner or pathway they do for employees.

NIOSH's framework offers the underpinning to integrate patient safety and employee safety and has been found to be a valid foundation for occupational health and safety research (McCaughey et al. 2013a; McCaughey, Turner, Kim, DelliFraine, and McGhan 2015). A recent review of the antecedents of health care employee injury validated the framework's pathways for preventing adverse safety events in the health care environment (McCaughey, Kimmel, Lukas, Walsh, Savage, and Halbesleben 2016).

Instances of adverse safety events are influenced by the "organization of work," which includes external forces, organizational practices, and work processes that all guide job design (Sauter et al. 2002). External factors are the economic, legal, and regulatory forces influencing how health care functions at the organizational level. Organizational practices are the structures and processes that lead and direct the organization. Work processes are the local job demands and conditions within a specific workplace. These factors are all aligned with the structures and processes driving improvements in patient safety climate.

Across industries, the emergence of management safety actions/leadership as a dominant dimension of employee safety climate perceptions is well established (Flin Mearns, O'Connor, and Bryden 2000; McCaughey, Halbesleben, Savage, Simons, and McGhan 2013b). Flin and colleagues' (2000) examination of the various safety climate scales utilized across industry sectors found perceptions of management behavior and attitudes toward safety to be the most commonly assessed factor in safety climate studies. A similar review of safety climate scales used specifically in health care also found management to be the most commonly assessed safety climate factor as well as being a significant positive factor in many outcomes (Flin, Burns, Mearns, Yule, and Robertson 2006). Patient safety is also optimized in work environments that value teamwork, have open and honest communication, support a blame-free environment, learn from previous adverse safety events, and utilize evidence-based information to drive best practices in all facets of care provision (Sammer et al. 2010).

Continuing with the NIOSH framework, another antecedent to adverse safety events is "safety and health services and programs." Safety and health services and programs include workplace safety programs, wellness initiatives, occupational training, teamwork efficiency training, conflict resolution training, leadership, and work/life balance initiatives. The organizational and work contexts influence specific job characteristics which may increase employees' exposure to psychological stress and physical hazards ultimately resulting in increased risk for adverse safety events for patients and employees (Sauter et al. 2002). These hazards can potentially be ameliorated by safety and health services and programs, which may moderate the risk and act as a resource allowing employees to deal with psychological stress and physical hazards more effectively. Given its multi-factorial approach, the NIOSH model provides a framework to evaluate, reduce or even prevent adverse safety event antecedents on multiple levels as well as assessing the impact of job design and managerial practices on patient and employee safety climate (McPhaul and Lipscomb 2004; Sauter et al. 2002).

Conclusion

The purpose of this integration of the patient safety and employee safety climate literatures was threefold. The first was to identify the industry impetus for the growing attention to patient safety climate while simultaneously examining the comparative inattention to employee safety climate. The second was to apply Flin's (2007) model as a guide for integrating the antecedents of patient safety climate and employee safety climate to propose a single model of employee behavior pathways that can impact the occurrence of adverse safety events for patients and employees. The final purpose was to apply the NIOSH (2002) framework to an integrated safety climate for health care organizations so optimal safety environments can be developed with the end goal of improving outcomes for patients and employees.

Reducing the excessive rates of adverse events experienced by care providers and patients is a critical need for health care organizations. Organizing safety climate studies around an evidence-based, conceptual framework offers managers and other leaders the opportunity to identify adverse safety event antecedents and address these with relevant safety programs with targeted prevention strategies. These initiatives can present significant challenges when designing and structuring work processes, determining staff and support needs based on unit specific work demands, and training organizational leaders to promote a positive and safe workplace for all. The integration of research on patient and employee safety climate may lead to innovative interventions designed to improve the organizational environment and reduce adverse safety events for care providers and their patients. One promising innovation is examining the dyadic relationship between care providers and patients and how that relationship impacts the safety climate. Synchronous integration of safety antecedents may be one avenue for achieving health care's goal of preventing patient adverse events while fostering a safer work environment for those providing the care.

Acknowledgments

The authors would like to gratefully acknowledge the invaluable editorial suggestions and directions from Dr. Tim Vogus; this is a better manuscript as a result.

References

Bureau of Labor Statistics. (2013). Nonfatal Occupational Injuries and Illnesses Requiring Days Away From Work, 2012. www.bls.gov/iif

Carr, J. Z., Schmidt, A. M., Ford, J. K., and DeShon, R. P. (2003). "Climate Perceptions Matter: A Meta-Analytic Path Analysis Relating Molar Climate, Cognitive and Affective States, and Individual Level Work Outcomes." *Journal of Applied Psychology*, 88(4), 605.

Clarke, S. (2006). "The Relationship Between Safety Climate and Safety Performance: A Meta-Analytic Review." *Journal of Occupational Health Psychology*, 11(4), 315-327.

Colley, S. K., Lincolne, J., and Neal, A. (2013). "An Examination of the Relationship Amongst Profiles of Perceived Organizational Values, Safety Climate and Safety Outcomes." *Safety Science*, 51(1), 69-76.

Danna, K., and Griffin, R. W. (1999). "Health and Well-Being in the Workplace: A Review and Synthesis of the Literature." *Journal of Management*, 25(3), 357-384.

Flin, R. (2007). "Measuring Safety Culture in Healthcare: A Case for Accurate Diagnosis." *Safety Science*, 45(6), 653-667.

Flin, R., Mearns, K., O'Connor, P., and Bryden, R. (2000). "Measuring Safety Climate: Identifying the Common Features." Safety Science, 34, 177–192.

Flin, R., Burns, C. Mearns, K., Yule, S., and Robertson, E.M. (2006). "Measuring Safety Climate in Health Care." *Quality and Safety in Health Care*, 15, 109–115.

Gandhi, T. (2015). The Changing Landscape of Patient Safety. Presentation at the MHA Patient Safety and Quality Symposium, Detroit, MI. www.mha.org/keystone_center/symposium/2015/documents/the_changing_landscape

Gomaa A. E., Tapp, L.C., Luckhaupt, S.E., Vanoli, K., Sarmiento, R.F., Raudabaugh, W.M., Nowlin, S., and Sprigg, S.M. (2015). "Occupational Traumatic Injuries among Workers in Health Care Facilities—United States, 2012–2014." Centers for Disease Control and Prevention: Morbidity and Mortality Weekly Report, Rep. 2015, 64: 405-410.

Health and Safety Commission Advisory Committee on the Safety of Nuclear Installations. (1993*). Organizing for Safety: Third Report of the ACSNI Study Group on Human Factors.* Sudbury, UK: HSE Books.

Institute of Medicine (IOM). 1999. *To Err Is Human: Building a Safer Health System.* Washington, DC: The National Academies Press.

Joint Commission Resources (Joint Commission International) (2010). Failure Mode and Effects Analysis in Health Care: Proactive Risk Reduction. www.jcrinc.com/assets/1/14/FMEA10_Sample_Pages.pdf

McCaughey, D., DelliFraine, J.L., McGhan, G.E., and Bruning, N.S. (2013a). "The Negative Effects of Workplace Injury and Illness on Workplace Safety Climate Perceptions and Health Care Worker Outcomes." *Safety Science* 51(1): 138–147.

McCaughey, D., Halbesleben, J., Savage G., Simons, T. and McGhan, G.E. (2013b). "Safety Leadership: Extending Workplace Safety Climate Best Practices across Health Care Workforces." *Advances in Health Care Management*, 14: 189-217.

McCaughey, D., Turner, N., Kim, J., DelliFraine, J.L., and McGhan G.E. (2015). "Examining Workplace Hazard Perceptions and Employee Outcomes in the Long-Term Care Industry." *Safety Science*, 78: 190-197.

McCaughey, D. Kimmel, A., Lukas, T., Walsh, E., Savage, G. and Halbesleben, J. (2016) "Antecedents to Workplace Injury in the Health Care Industry." *Health Care Management Review*, 41(1): 42-55.

McPhaul, K. M., and Lipscomb, J. A. (2004). "Workplace Violence in Health Care: Recognized but Not Regulated." *Online Journal of Issues in Nursing*, 9(3), 7.

National Quality Forum (NQF). (2011). *Serious Reportable Events in Healthcare—2011 Update: A Consensus Report.* Washington, DC: NQF.

Neal, A., Griffin, M.A., and Hart, P.M. (2000). "The Impact of Organizational Climate on Safety Climate and Individual Behavior." *Safety Science*, 34(1–3), 99–109.

Occupational Safety and Health Administration. (2011). Statement from Secretary of Labor Hilda L. Solis on Reported Decline in Workplace Injuries and Illnesses. www.dol.gov/opa/media/press/osha/OSHA20111547.htm

Occupational Safety and Health Administration. (no date). Organizational Safety Culture—Linking Patient and Worker Safety. www.osha.gov/SLTC/healthcarefacilities/safetyculture.html

Parker, C. P., Baltes, B. B., Young, S. A., Huff, J. W., Altmann, R. A., Lacost, H. A., and Roberts, J. E. (2003). "Relationships Between Psychological Climate Perceptions and Work Outcomes: A Meta-Analytic Review." *Journal of Organizational Behavior*, 24(4), 389-416.

Reiman, T. and Rollenhagen, C. (2014). "Does the Concept of Safety Culture Help or Hinder Systems Thinking in Safety?" *Accident Analysis and Prevention*, 68, 5-15.

Sammer, C.E., Lykens, K., Singh, K.P., Mains, D.A., and Lackan, N.A. (2010). "What Is Patient Safety Culture? A Review of the Literature." *The Journal of Nursing Scholarship*, 42:2, 156-165.

Sauter, S.L. et al., (2002). *The Changing Organization of Work and the Safety and Health of Working People.* Center for Disease Control and Prevention, Department of Health and Human Services. Cincinnati: NIOSH Publications Dissemination. www.cdc.gov/niosh/docs/2002-116

V. LERA Best Posters

Changes in the Power of the HR Executive in Japanese Firms 1990–2015

Tomoyuki Shimanuki
Hitotsubashi University, Japan

This research explores how the power of Japanese human resource (HR) executives has changed over the past 25 years. Our original panel data on the board members of Japanese large firms reveals that (1) the proportion of firms with HR executives has greatly decreased, although the proportion of firms with corporate planning (CP) and financial executives remains almost unchanged; (2) the proportions of firms with HR executives doubling as CP or financial executives have increased; and (3) the proportion of firms with corporate administrative executives supervising HR executives has increased. The power of current Japanese HR executives is more vulnerable and dependent on other functional executives than in the past.

Hard or Soft Forms of Conflict: Workers' Perception in the US, the UK, and Canada

Umar M. Boodoo
London School of Economics and Political Science

Lorenzo Frangi
University of Québec in Montréal

Rafael Gomez
University of Toronto

Robert P. HebdonS
McGill University

Private sector unions are in decline. Their future is dependent on engaging workers to the point of active participation. Previous research has demonstrated that such participation is preceded by the development of some level of member loyalty. Our study examines key antecedents to union loyalty and identifies a key mechanism through which these factors work. Specifically, we find pro-union attitudes and union instrumentality to be significant predictors of union member loyalty. Additionally, procedural justice perceptions mediate the relationship between antecedents and loyalty. Thus, our findings reflect the view that union socialization may be important in developing member loyalty.

VI. Occupational Regulation in the Age of Uber

Licensure or License[1]: Reconsidering Occupational Regulation

Dante DeAntonio
Robert Thornton
Lehigh University

Edward Timmons
St. Francis University

Introduction

The scope of occupational licensing in the U.S. has grown enormously in recent decades. For example, in the 1950s it was estimated that fewer than 5% of workers in the U.S. needed a license to work at their jobs. But by 2008, this figure was close to 30%. In 2003 CLEAR estimated that more than 800 different occupations were licensed in at least one state, while today more than 1,100 occupations are said to be subject to licensing, certification, or registration requirements,[2] although both the occupations and their numbers vary greatly across the states (Council of State Governments; Kleiner and Krueger 2013; CLEAR).

Until fairly recently, occupational licensing was a subject that was largely ignored—not just by labor economists, but by the general public as well. The pros and cons of requiring a license to practice certain occupations were not much talked about and certainly not subject to much criticism. After all, licensing could be said to protect consumers from incompetent or disreputable practitioners (think doctors here). It could also be said to ensure a high level of quality of service (dieticians and barbers, for example). As of late, though, occupational licensing has begun to attract a growing and increasingly vocal stream of critics. The criticisms reflect a number of concerns. For example, it has been alleged that too many occupations now require a license to practice and that the requirements for attaining a license in terms of costs and length of training are often excessive. Complaints have also been made that higher prices for the services provided by licensed practitioners are too often the result and that the benefits to consumers in terms of higher quality are sometimes nonexistent. Finally, it has been charged that excessive licensing has resulted in adverse effects on employment opportunities and worker mobility, especially for those with lower levels of education.

In light of this new and growing criticism of excessive occupational licensing, what are the prospects for deregulation—or what we shall refer to as "de-licensing"? This is the subject of our paper. In the next section, we discuss in more detail the reasons for the growing concern and criticisms that occupational licensing is facing. We then analyze the efforts that nine states to date have taken in the past four years to deregulate (de-license) certain groups of occupations.

These de-licensing proposals have generally not gone through the usual sunset review process. The sunset review process mandated in many states involves periodic assessments of licensed occupations and licensing boards as well as their possible termination unless continued by the legislature. Finally, we examine whether there are any particular state characteristics—economic, demographic, or political—that seem to increase the likelihood of de-licensing activity or proposals. We know of no theories of de-licensing per se, but there is a small extant literature concerning the various factors that tend to be associated with the *passage* of licensing legislation. Could some of these same factors explain recent attempts to de-license occupations?

Author's address: Lehigh University, College of Business and Economics, RBC 460, 621 Taylor Street, Bethlehem, PA 18015

The Growing Outcry Against Occupational Licensing

Although the "case" against occupational licensing was first made popular by Milton Friedman (1962) more than 50 years ago, the hue and cry against it did not gather much popular support until the last decade.[3] The reasons for the turnaround are several. First, licensing has spread into occupations for which the protection of the public hardly seems necessary and, in some cases, where requiring a license to practice seems ludicrous. For example, in a list of what he refers to as the nation's "most outrageous licensing laws," Adam Summers notes that fortune tellers require a license in Maryland, junkyard dealers in Ohio, rainmakers in Arizona, and manure applicators in Iowa (Summers 2007, p. 43). Furthermore, yoga instructors are licensed in more than a dozen states, while eyebrow threaders and art therapists are among some newly licensed occupations.

The long-accepted argument that licensing raises the quality of services provided to the public has not in fact been strongly supported by the evidence. As Morris Kleiner has found, "Overall few studies have shown significant benefits of occupational regulation on the quality of service received by consumers …" (Kleiner 2015, p. 13).

There is a growing realization that, by restricting entry into occupations through the requirement of a license to practice, the result can be fewer employment opportunities (especially for those with lower levels of education) as well as higher prices for goods and services. This has been the motivation for several recent state legislative proposals to deregulate certain occupations.

Finally, calls for licensing—as well as resistance to de-licensing—almost never originate from consumers, but rather from practitioners in the occupation itself, who often see licensing as a means to create economic rents by protecting themselves from competitors. Moreover, in virtually all cases when an occupation is newly licensed, existing practitioners are exempt from the licensing requirements by grandfather clauses.

The recent stream of criticism alluded to earlier has emanated from many channels: economists, the news media, and various state governments. In the case of economists, for example, as Morris Kleiner noted as late as 2000:

> [E]ven though occupational licensing has historically been among the most examined institutions in labor economics [see, for example, Adam Smith] this institution has received relatively little *recent* [present authors' italics] attention, either from academics or the public policy press. An examination of the *American Economic Review*, *Journal of Political Economy* and the *Quarterly Journal of Economics* found no articles published in these journals on occupational licensing during the past five years. (p. 190)

However, over the 15 years since Kleiner's observation, our search of EconLit has uncovered at least 32 economics articles or books published since then with "occupational licensing" in the title and more than 3,100 text mentions of the term. A search of Google Scholar, which casts a much wider net than EconLit because it searches scholarly literature across all disciplines, reveals 112 published works with "occupational licensing" in the title, and more than 65,000 text mentions of the term since 2000.

The news media have also not been reluctant to unleash criticism of certain licensed occupations and the excessive costs and restrictions that licensing often requires for one to become licensed. For example:

- A 2014 article in *The New Republic* asks, "Does a 'Shampooer' Really Need 70 Days of Training?" (Vinik 2014).
- In "Practicing Unlicensed Geology" (*The Ledger,* December 20, 2007), it was reported that a man at a public hearing in Florida who spoke out against a proposed sand-mining operation was slapped with a "cease and desist" court order for practicing geology without the required license.
- The news article "Court Says State Law Has Teeth" (Barron 2007) related the situation of a *manufacturer* of dentures who was forced to quit his practice because the Wyoming high court ruled he was practicing dentistry without a license.
- "It's Illegal for Monks to Sell Caskets in Louisiana," reported *Bloomberg Business Week* (June 1, 2012). Until recently, only licensed funeral establishments could sell caskets in Louisiana.

- In "Louisiana Prunes Thorny Licensing Law," Tresa Baldas (2010) explains how Louisiana abolished a law that had required aspiring florists to pass a 4-hour-long floral arranging demonstration (exam) before they could be given a mandatory florist license. The demonstration was given before a panel consisting of (guess who?) licensed florists, who were obviously potential competitors.
- "So You Think You Can Be a Hair Braider?" asks a *New York Times* article (Goldstein 2012) that points out that an African hair braider could not practice her profession in Utah without a cosmetology license. The license would have required $16,000 in tuition for about two years of cosmetology school training—training that did not even include African hair braiding.

In light of the growing criticism of excessive occupational licensing, what are the prospects for "de-licensing"? And through what channels could effective de-licensing take place? If history alone is any indication, the prospects for widespread de-licensing would not seem to be very promising. To explain further, our recent research (Thornton and Timmons 2015) on occupational deregulation attempts over the past 40 years has uncovered only eight cases of an occupation licensed at the state level being de-licensed by legislative action (none all that recent). They are:

- Barbers in Alabama (1983)
- Morticians in Colorado (1971)
- Naturopaths in Virginia (1972)
- Private investigators in Colorado (1977)
- Egg candlers in Colorado (1994)
- Interior designers in Alabama (2004)
- Watchmakers in Minnesota (1983) and in Wisconsin (1979).

What is interesting is that in half of these cases attempts to re-license the occupations followed soon afterward (perhaps reminding one of the popular "Whack-a-mole" arcade game?) And in one case, the attempt was successful, with the practice of barbering once again (as of 2014) requiring a license in Alabama.[4]

Furthermore, as noted earlier, most states have a sunset review process that involves periodic reviews (usually called "legislative audits" or "performance audits") of licensing and licensing boards and their possible termination unless continued by the legislature. In theory, the legislature's decision to terminate or continue is based on the sunset review panel's recommendation. But in fact, these audits nearly always recommend the continuation of licensure. And even when recommendations are made to remove licensing, the legislature generally ignores the recommendation. Strong special interests are often very effective in resisting calls to de-regulate a particular licensed occupation. After all, practitioners potentially have much more to lose should their profession be de-licensed than consumers (on an individual basis) have to gain. A good example is the case of cosmetologists, whose professional association (the PBA, Professional Beauty Association) has fought hard against deregulation of their profession. For example, in 2012 a bill was introduced in the Indiana General Assembly that would have eliminated mandatory licensing for cosmetologists, as well as for a number of other occupations. However, only a week after the bill was introduced, it was withdrawn by its sponsor. The reason given was the loud public outcry opposing the bill, coming mainly from cosmetologists. More than many other professional associations, the cosmetologists have been very aggressive in their attempts to ward off de-licensing. On the PBA website, the PBA Director of Government Affairs warns members to "Beware the 'D' word: occupational licensing under attack." The warning further states that if the deregulation of cosmetology were to come about, anyone with no formal training would be able to practice cosmetology, thus putting consumers at risk of injuries, burns, infections, and the spread of diseases, such as hepatitis and methicillin-resistant staphylococcus, due to unsanitary practices. The PBA director also advises members to "Stand up for your profession! You have the knowledge and power to speak out against licensing proponents [sic: we presume that she means *opponents*] and educate legislators in your state about the importance of education and the true risks consumers face without oversight of this hands-on industry…."[5]

Despite these obstacles, new avenues leading to potential deregulation have recently emerged. President Obama has spoken out against excessive licensing as a "job-killer" and in his FY 2016 budget included $15 million in funding at the Department of Labor to "identify, explore, and address areas where licensing requirements create barriers to labor market entry or labor mobility." ("Occupational Licensing: A Framework for Policymakers," p. 41). The problem, however, is that the licensing of occupations in the vast majority of cases is done at the *state* level. The federal government has little control. Nonetheless, in a recent joint report of the Department of the Treasury, the Department of Labor, and the Council of Economic Advisers ("Occupational Licensing: A Framework for Policymakers"), the substantial costs that occupational licensing imposes on workers and consumers were laid out, along with best practice recommendations to help ensure that occupational regulation might continue to protect consumers without "placing unnecessary restrictions on employment, innovation, or access to important goods and services" (p. 3).

Moreover, despite the fact that many professional associations (of which the PBA is only one example)[6] have taken up the hue and cry against occupational de-licensing, there is one institution in particular that has taken aggressive steps to push for occupational de-regulation. The Institute for Justice (IJ), which describes itself as a national law firm for liberty, has served as a pro bono advocate in numerous lawsuits filed by individuals restricted from practicing their trade or profession because of what it deems as overly strict licensing laws (e.g., hair braiders). The IJ has also published several research reports analyzing (and decrying) what it calls excessive occupational regulation, as well as "license creep." The latter refers to the expansion of definitional boundaries of some occupations, as in the case of eyebrow threaders being required to secure cosmetology licenses and horse-teeth filers being required to have veterinary licenses, for example (Carpenter et al., p. 32). The IJ's work has also received much favorable publicity (Bergal 2015).

Finally, and again very recently, attempts have arisen in a number of states to de-license *groups* of occupations rather than rely on the sunset process, which has turned out to be slow, costly, and largely ineffective. This is the subject to which we turn next. Where have such attempts arisen? Have they been successful? Where are such attempts likely to arise?

Recent State Proposals to Collectively De-license Certain Occupations

Since the year 2011, nine states have formulated legislative or administrative proposals dealing with occupational deregulation, including de-licensing. Table 1 contains summary information concerning the states, dates of the proposals, descriptions, and current status (see Thornton and Timmons 2015 for more discussion of these nine proposals). As can be seen from the table, the proposals generally share several common features.

Most of the state proposals (those of Florida, Indiana, Michigan, Missouri, New Hampshire, and Texas) would eliminate the licensing of a number (usually one or two dozen) occupations. The occupations suggested for de-licensing (arguably) do not concern public health or safety, and several (e.g., barbering and cosmetology) are proposed for deregulation in all of the state proposals. Many of the proposals have been predicated on the argument that unnecessary licensing restricts job creation and/ or opportunities for the disadvantaged. In a few cases (e.g., Missouri, Indiana, and Texas), proposals have been offered for stiffening requirements for future licensing.

The most striking observation from Table 1, however, is that the deregulation proposals to date have largely been unsuccessful. In most cases, the bills have either not been acted upon, have died in committee, or have been withdrawn because of political pressure. In most of the nine states (except for Connecticut), follow-up versions of the unsuccessful bills have later been advanced, only to meet similar fates.

TABLE 1
Recent Legislative or Administrative De-Licensing Proposals

State	Date(s)	Proposal	Description	Status
Connecticut	2013	SB324: An act requiring the Commissioner of Consumer Protection to undertake a study of occupational licenses.	Purpose is to recommend elimination of licenses for occupations where public health or safety is not an issue, as well as to eliminate unnecessary regulatory burdens on individuals and small businesses.	Has not been voted upon.
Florida	2011	A 2011 bill was passed by the Florida House that would deregulate 14 licensed occupations.	The occupations proposed for deregulation included auctioneers, athlete agents, hair braiders, interior designers, and fundraising consultants and solicitors. The bill encountered substantial industry opposition.	The bill was not passed by the Florida Senate despite the support of the governor.
Florida	2013	In 2013 bills similar to the 2011 bill (titled Deregulation of Professions and Occupations) were introduced in the Florida House and Senate (HB1189 and SB 720).		Both bills died in committee.
Indiana	2012	HB1006	Would eliminate mandatory licensing of barbers, cosmetologists, dietitians, hearing aid dealers, private investigators, and security guards.	Withdrawn after one week due to loud public outcry.
Indiana	2013	SB520	Would create a committee to "eliminate, reduce, and streamline employee regulation" [aka the "Eraser Committee"] and would eliminate mandatory licensing of 14 occupations over a 5-year period.	Bill failed to receive a hearing in the House.
Indiana	2014	SEA421	Act effectively encouraged occupations seeking new regulations to consider certification rather than licensing.	Act signed into law in March 2014.
Indiana	2014	Report issued by Jobs Creation Committee	Indiana General Assembly established the Jobs Creation Committee to assess licensing effectiveness. Report issued on 7/1/15 recommended de-licensing of several occupations.	No legislative action to date based upon report recommendations.
Michigan	2012	Several bills (including HB4688 Public Act 267 of 2014) that de-licensed dietitians	Michigan Office of Regulatory Reinvention (ORR) recommended the deregulation of 18 occupations (including dietician, forester, oculist, and polygraph examiner). Not all of the 18 occupations are licensed.	Several bills were passed in 2013 and 2014 deregulating occupations. Dietitians were
Minnesota	2012	HF2002, SF1629, "Licensing Relief and Job Creation Act"	Bill would allow a person practicing without a license in an occupation requiring a license to challenge the licensing requirement in court.	Bill was referred to committee and never voted on.

Minnesota	2015	A nearly identical bill (SF784) to the 2012 bill was introduced		Referred to committee: no action taken.
Missouri	2013	HB590	Bill would allow persons to practice the professions of interior design, barbering, and cosmetology without having to secure a license.	Referred to committee.
Missouri	2014	HB1891	Bill similar to HB590 expanded the number of professions mentioned in HB590 to 12, including massage therapists, embalmers, and athletic agents.	Died in committee.
Missouri	2014	HB1824	Bill would restrict the imposition of licensing requirements on occupations that were not regulated as of January 1, 2015. Principles were formulated to guarantee that individuals may engage in occupations of their choice "free from	Bill failed to receive votes to advance to floor.
New Hampshire	2011	HB446 (Repealing the Authority for Regulation of Certain Professional Occupations)	Bill would repeal the licensing of more than a dozen licensed occupations, including barbers, cosmetologists, massage therapists, hunting and fishing guides, and court reporters.	Bill was defeated in 2012.
New Hampshire	2012	HB1265 (Relative to Criteria for the Government Regulation of Occupations and Professions)	Bill would establish criteria for regulation of occupations and professionals by boards and commissions. Bill would also support certification ("volunteering licensing") rather than mandatory licensing.	In 2012, a legislative study committee recommended against advancing the bill to the legislature.
North Carolina	2011	HB587 (An Act to Promote N.C. Job Growth through Regulatory Reform)	Bill would create a study commission on occupational licensing to identify outdated and unnecessary occupational licensing laws that should be repealed as well as to study effective alternatives to occupational licensing laws.	Portions of the bill were incorporated into a Senate bill that was passed, but the provisions related to
Texas	2013	HB86 (Relating to the Criteria for Review by the Sunset Advisory Commission of an Agency That Licenses an Occupation)	Bill was not designed to de-license specific occupations, but rather to provide the Texas Sunset Advisory Commission with a broader set of criteria to be considered in continuing to license an occupation.	Bill was signed into law in September 2013.
Texas	2014	Staff report from Sunset Advisory Commission	Staff report recommended the de-licensing of 6 medical professions (e.g., dieticians, radiologic technologists, and perfusionists).	Final report to Texas state legislative removed the de- licensing recommendations.

As of the time of this writing, in only three states—Indiana, Texas, and Michigan—has there occurred what could be judged to be a limited success in occupational deregulation. In Indiana and Texas, several occupations have been deregulated (hypnotists and environmental health specialists in Indiana and opticians in Texas), but none of these occupations had previously been licensed. Rather they had been subject to less restrictive forms of regulation—registration or certification. Michigan has enjoyed somewhat more success. In all, seven occupations have been deregulated in Michigan in the last two years—auctioneers, community planners, dieticians/nutritionists, immigration clerical assistants, interior designers, oculists, and proprietary school solicitors. However, only dieticians/nutritionists had previously been licensed. The other six occupations had been subject to registration or certification requirements. Furthermore, another nine occupations (and/or their regulatory boards) have been recommended for deregulation (e.g., acupuncturist, polygraph examiners, occupational therapist, landscape architect), but no legislative action has yet been taken.

Explaining the Emergence of Deregulation Proposals[7]

Do these nine recent statewide attempts at collective de-licensing discussed above share any common characteristics? Other than the fact that most de-licensing attempts have taken place in the eastern United States (see states with shading in Figure 1), are there any other particular state characteristics—economic, demographic, or political—that seem to increase the likelihood of de-licensing activity? To answer this question, it would be useful to have a theory of deregulation or de- licensing. We know of no such theories, but there is a small extant literature that we briefly summarize below concerning the various factors that tend to be associated with the *passage*—the production—of licensing legislation. Could some of these same factors (or their absence) also explain attempts to de-license occupations?

In his classic article "The Theory of Economic Regulation," George Stigler (1971) used occupational licensing as an example of the use of the political process by practitioners of an occupation to benefit the group. He considered several characteristics of an occupation which he argued should affect the ability of the group to gain the necessary political power, including the size of the occupation (the larger the group, the more votes it has) and an urbanization measure. His regressions for a small sample of select occupations show only modest support for his theory, however.

FIGURE 1
States with Recent De-Licensing Proposals (as indicated by shading)

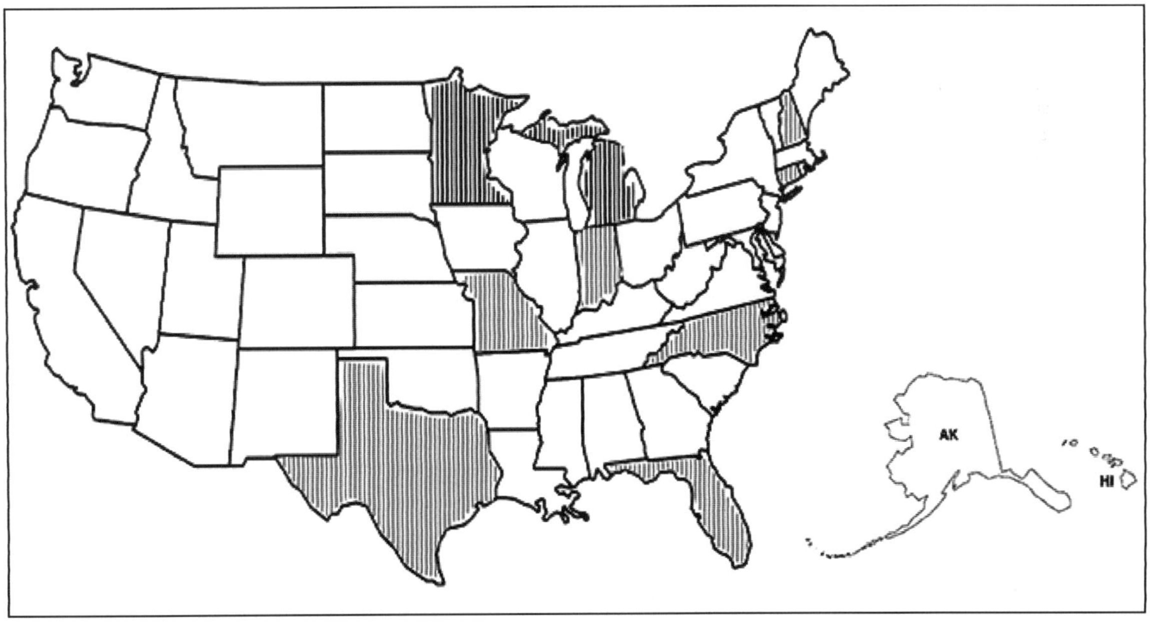

Smith (1982) empirically examined the extent to which changes in state occupational licensing laws reflected various political factors and the structure of state legislatures. Her explanatory variables include party concentration (the percentage of the state legislature belonging to the majority party), the number of constituents per legislator (the more constituents represented, the less costly for a legislator to support special interest legislation), and state per capita income. She finds that these and other factors are important for explaining the emergence of licensing laws over the period studied (1952-68).

Graddy (1991) analyzed changes in the number of states regulating five occupations (geologist, landscape architect, librarian, physician assistant, and psychologist) over the period 1968-1980. Her explanatory variables include various legislative variables (e.g., the proportion of the state legislature held by the majority party), various interest group variables (e.g., the number of professionals who are members of the major professional association), and various public interest variables. Graddy's logit regression results show that all three types of variables may be important in determining whether these occupations are regulated in the various states.

Could the same or similar political and economic variables that have been found to partially explain the passage of licensing legislation be also associated with the emergence of legislative proposals (albeit so far mostly unsuccessful) for de-licensing occupations? To analyze this question, we have estimated several regression equations, with the emergence of a bill to de- license over the years 2011-14 as the dependent variable (1=yes, as for the nine states discussed in the previous section, with some states having multiple attempts at de-licensing in that period). The state-characteristic independent variables that we have utilized are the following:

1. State per capita income
2. The number of low-income occupations that are currently licensed by the state
3. The state unemployment rate
4. The percentage of the state population that is minority
5. The percentage of the state population with a bachelor's degree or higher
6. Union density in the state
7. Whether the state is a right-to-work state
8. The number of constituents per legislator
9. A dummy variable indicating whether there was a Republican majority in the state Senate
10. The percentage of state legislators who are Republican

Table 2 presents summary statistics for each of these variables in the states for 2011–2014. The intuition behind most of the variables above is fairly straightforward: that both political and economic factors may also play a role in attempts to de-license occupations, just as the literature dealing with the determinants of the passage of licensing summarized above shows. For example, we have seen that the rationale offered for several of the state bills has been that excessive licensing laws have inhibited job growth (especially among minorities). Because practitioners of occupations that are licensed by the states are often members of professional associations or even unions (e.g., cosmetologists, nurses, plumbers) we would expect a negative association between de-licensing efforts and union density (with the reverse relationship expected in right-to-work states). For the variable specified as the percentage of low-income occupations that are currently licensed by the state, we have used a measure formulated by Carpenter et al. (2012) to gauge the "burden of licensure" for lower-income workers.

TABLE 2
Summary Statistics

Variable	Obs	Mean	Std. Dev.	Min	Max
Proposed Legislation	200	0.07	0.26	0	1
Right-to-Work State	200	0.47	0.50	0	1
Unemployment Rate	200	6.99	1.83	2.80	13.10
Per Capita Income ($000's)	200	43.48	6.55	32.11	62.47
Pct. Minority Population	200	19.91	12.20	4.53	74.13
Pct. College Degree	200	28.65	4.88	18.50	40.92
Union Density	200	10.46	5.34	1.90	24.60
Pct. Low-Income Occupations Licensed	200	42.63	10.44	23.53	69.61
Constituents per Legislator (000s)	200	43.26	49.31	3.11	323.35
Republican Senate Majority	200	0.57	0.50	0	1
Republican House Majority	200	0.59	0.49	0	1
Overall Pct. Republican	200	52.02	18.78	0	86.67

Data sources: Bureau of Economic Analysis; Institute for Justice; Bureau of Labor Statistics; Bureau of the Census; American Community Survey; Current Population Survey; American Community Survey; National Conference of State Legislatures; Council of State Governments.

Specifically, the variable looks at 102 occupations that are licensed in at least one state and that are recognized by the U.S. Bureau of Labor Statistics as ones in which practitioners' earnings are lower than the national average. The more such occupations that are licensed in a state, arguably the stronger the case that might be made for de-licensing.

To explore further the characteristics of these states, we estimate a probit model using the variables described above, along with time fixed effects and geographic fixed effects (in some cases). The effects of these variables on the likelihood of the proposal of a legislative de-licensing bill (hereafter *Proposal)* are displayed in Table 3. The model results in column 1 include only time fixed effects, column 2 includes time fixed effects and controls for census regions, and column 3 includes time fixed effects and controls for census divisions. Due to the small number of states proposing legislation, using either set of geographic controls causes the loss of some observations when there is no variation in the dependent variables within a given region or division.

In the regression results in columns 1, 2, and 3 respectively, there are statistically significant associations between the *Proposal* variable and the state's unemployment rate (positive), per capita income (positive), percent minority population (negative), and (excluding column 3) percent college degree (positive). The Republican Senate majority and Republican House majority variables are positive and statistically significant across all three versions of the model also. Our results seem to suggest that the composition of the House is more important in predicting the likelihood of a de-licensing proposal. A Republican majority in a state's House is associated with a 25 percentage point increase in the likelihood of a de-licensing proposal with a Republican majority in the Senate increasing the likelihood by as much as 15 percentage points. These results are not surprising, however, in light of the fact that it has been Republican legislators who have proposed most of the de-licensing bills in the nine states analyzed.

TABLE 3
Determinants of State De-Licensing Proposals: Probit Model Estimates

	(1)	(2)	(3)
VARIABLES	**Proposed Legislation**	**Proposed Legislation**	**Proposed Legislation**
Right-to-Work State	0.220 (0.617) [0.0202]	0.333 (0.714) [0.0299]	0.349 (0.681) [0.0366]
Unemployment Rate	0.842*** (0.261) [0.0770]	2.188*** (0.467) [0.195]	2.139*** (0.524) [0.224]
Per Capita Income	0.0982* (0.0509) [0.0090]	0.439*** (0.120) [0.0390]	0.453*** (0.124) [0.0474]
Pct. Minority Population	-0.0786*** (0.0234) [-0.0072]	-0.273*** (0.0657) [-0.0243]	-0.270*** (0.0995) [-0.0282]
Pct. College Degree	0.168** (0.0687) [0.0153]	0.232** (0.104) [0.0206]	0.130 (0.112) [0.0136]
Union Density	-0.0344 (0.0593) [-0.0031]	-0.0148 (0.125) [-0.0013]	0.0755 (0.137) [0.0079]
Pct. Low-Income Occupations Licensed	-4.703** (1.861) [-0.0043]	-15.89*** (4.643) [-0.0141]	-15.79*** (5.553) [-0.0165]
Constituents per Legislator	0.0039 (0.0038) [0.0004]	-0.0133 (0.0083) [-0.0018]	-0.0143 (0.0113) [-0.0015]
Republican Senate Majority	2.281*** (0.822) [0.1331]	1.722** (0.807) [0.109]	2.034** (0.885) [0.147]
Republican House Majority	1.522** (0.634) [0.0983]	5.075*** (1.585) [0.249]	4.678*** (1.738) [0.226]
Overall Pct. Republican	0.274 (2.454) [0.0003]	2.422 (3.861) [0.0022]	2.927 (3.867) [0.0031]
Constant	-15.99*** (4.906)	-34.31*** (7.777)	-32.62*** (7.922)
Census Region Controls	N	Y	N
Census Division Controls	N	N	Y
Observations	200	148	120

Robust standard errors in parentheses.
Marginal effects in brackets.
*** $p<0.01$, ** $p<0.05$, * $p<0.1$

We must emphasize that our analysis is simply a heuristic first look at some possible common characteristics shared by states that have seen legislative de-licensing proposals advanced within the last several years. Our results do not necessarily imply causal relationships between the composition of a State's House and Senate and its likelihood of a *Proposal*, or causal relationships between *Proposal* and other independent variables for that matter.

Conclusion

Where does this leave us? As we have seen, occupational licensing in the U.S. has grown rapidly in the past several decades. Over most of this period, there has been little concern over, and not much attention devoted to, this phenomenon. Most recently, however, concerns have arisen about the extent, the costs, and the job-killing nature associated with the licensing of many occupations. Despite this recent attention, though, the prospects for widespread de-licensing in the immediate future appear to be slim. First of all, past examples of successful de-licensing have been few—a total of nine (counting Michigan's recent de-licensing of dieticians) in the span of 40 years! If any widespread de-licensing is to take place, it will not likely occur through the sunset review process that operates in most states, which has proved to be slow, costly, and ineffective. Instead, such de-licensing is more likely to occur through state initiatives to de-license *groups* of occupations. Such initiatives, as we have seen, are more likely to arise in states with Republican majorities in the state House or Senate, as well as in states with higher unemployment rates, higher per capita incomes, and a higher proportion of college graduates. Still, these attempts have not so far resulted in a large number of occupations being deregulated, much less de-licensed.

Perhaps a recent comment from the executive director of CLEAR (Adam Parfitt) best summarizes the issue. According to Parfitt, "I think the atmosphere [concerning licensing] has changed in tone. Whether that's translating yet into widespread deregulation, I'd say probably not" (CLEAR, November 12, 2015).

Endnotes

[1] "License" 3(a): Freedom that allows or is used with irresponsibility (*Webster's Ninth New Collegiate Dictionary*).

[2] Licensing laws restrict the practice of an occupation to those who hold a license while certification laws restrict the use of the title, but not the practice, to those who are certified. Registration requirements merely stipulate that individuals practicing a certain occupation must list their names on some official register.

[3] In the 1980s and early 1990s, the Federal Trade Commission examined the issue of occupational licensing (e.g., Cox and Foster 1990), and the American Enterprise Institute (Rottenberg 1980) organized a conference on the topic. None of these activities resulted in significant legislative attempts to scale back occupational regulation, however.

[4] The Whack-A-Mole arcade game consists of a flat surface with several round holes. Each hole contains a single plastic mole and the machinery to move it up and down. Once the game starts, the moles keep popping up from their holes at random. The object of the game is to force the individual moles back into their holes by hitting them directly on their heads with a large soft mallet. The analogy to repeated re-licensing attempts is, we think, both apt and humorous.

[4] Myra Irizarry, "Beware the 'D' Word," Professional Beauty Association. The PBA also claims to have conducted a poll of 1,200 American voters in 2012, with the results showing that "more than nine in ten (94%) voters say that they support requiring their stylist, barber, nail technician or esthetician to be licensed."

[6] Respiratory therapist associations in Texas are also advocating against de-licensing of the profession rtfocus.com/signpetition/

[7] These recent statewide collective attempts are not the first to have been tried. In 1977, the Georgia legislature passed a bill ("An Act Providing for the Review, Continuation, Reestablishment, or Termination of Regulatory Agencies") that set termination dates for a number of licensing boards. But what ultimately happened was that, as the termination dates drew near, the legislature passed bills to halt the termination. For example, in 1989 Georgia scheduled the elimination of the licensing board for dieticians to take place in 1995, but the Georgia legislature then halted the elimination the year before it was to occur.

References

Baldas, Tresa. 2010. "Louisiana Prunes Thorny Licensing Law," *National Law Journal*, July 12.

Barron, Joan. 2007. "Court Says State Law Has Teeth" *Star-Tribune*, August 18.

Bergal, Jenni. 2015. "A License to Braid Hair? Critics Say State Licensing Rules Have Gone Too Far." The Pew Charitable Trusts. www.pewtrusts.org/en/research-and-analysis/blogs/stateline/2015/1/30/a-license-to-braid-hair-critics-say-state-licensing-rules-have-gone-too-far

Carpenter, Dick, Lisa Knepper, Angela Erickson, and John Ross. 2012. *License to Work*. Arlington, VA: Institute for Justice.

Council of State Governments. 1952. *Occupational Licensing Legislation in the States*, Chicago: Council of State Governments.

Cox, Carolyn, and Susan Foster. 1990. "The Costs and Benefits of Occupational Regulation." Federal Trade Commission Report, www.ftc.gov/system/files/documents/reports/costs-benefits-occupational-regulation/cox_foster_-_occupational_licensing.pdf

Friedman, Milton. 1962. "Occupational Licensing," in *Capitalism and Freedom*, Chicago: University of Chicago Press.

Graddy, Elizabeth. 1991. "Toward a General Theory of Occupational Regulation." *Social Science Quarterly*. Vol. 72, pp. 676-95.

Goldstein, Jacob. 2012. "So You Think You Can Be a Hair Braider?" *New York Times*, June 12.

Irizarry, Myra, "Beware the 'D' Word: Cosmetology Licensing under Attack," Professional Beauty Association.

"It's Illegal for Monks to Sell Caskets in Louisiana." 2012. *Bloomberg Business Week*, June 1.

Kleiner, Morris. 2015. "Reforming Occupational Licensing Policies." The Hamilton Project. Brookings Paper 2015-1, March.

Kleiner, Morris. 2000. "Occupational Licensing," *Journal of Economic Perspectives*, Vol. 14, No. 4, pp. 189-202.

Kleiner, Morris, and Alan Krueger. 2013. "Analyzing the Extent and Influence of Occupational Licensing on the Labor Market." *Journal of Labor Economics* Vol. 31, pp. S173-S202.

"Occupational Licensing: A Framework for Policymakers." 2015. U.S. Department of the Treasury, U.S. Department of Labor, and Council of Economic Advisers, July.

"Practicing Unlicensed Geology" [also titled "Bureaucratic Madness"]. 2007. *The Ledger*, December 20.

Rottenberg, Simon. (Ed.). 1980. *Occupational Licensure and Regulation*. Washington, DC: American Enterprise Institute.

Stigler, George. 1971. "The Theory of Economic Regulation." *The Bell Journal of Economics and Management Science*, Vol. 2, pp. 3-21.

Summers, Adam. 2007. *Occupational Licensing: Ranking the States and Exploring Alternatives*, The Reason Foundation.

Vinik, Danny. 2014. "Does a 'Shampooer' Really Need 70 Days of Training?" *The New Republic*, May 22.

Smith, Janet K. 1982. "Production of Licensing Legislation: An Economic Analysis of Interstate Differences," *Journal of Legal Studies*, Vol. 11, pp. 117-37.

Thornton, Robert, and Edward Timmons. 2015. "The De-licensing of Occupations in the US," *Monthly Labor Review*.

"The Washington Post Comments on Deregulation Initiatives." 2015. CLEAR News, November 12.

VII. Distraction or Catalyst? Trade Agreements and Labor Standards

The Meaning of Labor Clauses in Free Trade Agreements

Janice R. Bellace
Wharton School
University of Pennsylvania

The current debates on the Trans-Pacific Partnership (TPP) present diametrically opposed views on the impact of liberalizing trade. Union resistance to free trade agreements (FTAs) for the most part focuses on job losses. FTAs per se are not responsible for this; rather, American jobs are lost to lesser-developed countries because wages and working conditions are much lower. Labor clauses have been inserted in FTAs as a sign of the signatories' commitment to worker rights and fair labor standards. But labor clauses in FTAs have not been successful in meeting this concern. Despite being cast as the "highest" labor standards in an FTA, the TPP fails to set forth fundamental worker rights with any specificity and leaves the setting of "appropriate" working conditions to each country to decide. The current text of Chapter 19 of the TPP and in particular its footnotes result in worker rights that have little meaning and place a heavy burden of proof on those seeking to prove a violation. Without a stronger and more specific labor rights clause, there is little basis for worker rights advocates to raise a successful complaint, especially in light of the historically weak and slow dispute resolution process in American FTAs.

Introduction

After a decade of negotiations, the governments of 12 nations[1] signed the Trans-Pacific Partnership (TPP), a free trade agreement (FTA) on February 4, 2016. It now must go through the internal ratification process of each country within two years.[2] The TPP has generated much controversy, particularly during a presidential election year. The amount of controversy in some ways is surprising as the United States already has trade agreements with half of these countries.[3] In part, the criticism of the TPP reflects many Americans' growing frustration with the perceived impact of FTAs on jobs domestically. In part, it reflects the fact that the TPP is viewed as a harbinger of the future. As the Obama administration has already observed, more countries in future may join the TPP, thus raising the likelihood that its importance will be even greater than the impact in the 12 signatory countries. The TPP may also serve as the template for future trade pacts because, as U.S. Trade Representative Michael Froman says, the TPP "will set the rules of the road."[4]

The TPP is also viewed as affecting negotiations on another FTA. At the same time, the United States has been negotiating an FTA with the European Union (EU), called the Transatlantic Trade and Investment Partnership (TTIP) although progress has been slow. The trade ties between the United States and European Union member states are extensive and, in general, there are few disputes. Tariff barriers are already quite low (on average under 3%), such that the main attraction of the TTIP is the removal of non-tariff barriers. For the United States, the TTIP is unusual in that the United States for the first time is negotiating with an entity as large, as advanced and as wealthy as it is. The negotiations have been difficult because powerful companies in a given industry in different countries

Author's address: Wharton School, 672 Jon M. Huntsman Hall, 3730 Walnut Street, Philadelphia, PA 19104 may hold sharply different views on issues such as what are appropriate automobile vehicle safety regulations as opposed to unnecessary regulatory barriers. This difference in opinion on the two sides of the Atlantic arises in other areas. In the European Union, unions are much stronger than in the United States and pro-labor views are advanced by social democratic parties. This has generated another area of tension in the TTIP negotiations since the labor clause in the FTAs the European Union has negotiated with other countries is stronger and more specific than the labor clause in American FTAs in that all EU FTAs expressly link worker rights to the ILO's eight core conventions (Agusti-Panareda, Ebert, and LeClerq 2015). European unions fear the loss of jobs, as more European Union companies may source goods and services from the United States, and in particular from certain states, where the rate of unionization is negligible and where wage costs and social benefits are much lower than in the European Union.

Responding to critics, U.S. Trade Representative Michael Froman has stated that the TPP "raises labor and environmental standards around the world to the highest level ever, and these are fully enforceable standards" (Council on Foreign Relations 2016). Froman may be technically correct in saying that the TPP has the highest level labor standards ever in an American FTA, but that is quite different from its being a "high standard" clause. Moreover, the assertion that it is fully enforceable is highly dubious as crucial rights are not only not defined but the TPP expressly severs the link to international instruments defining these rights. It is difficult to argue that the TPP's labor clause has real meaning because it is impossible to identify definitions of these rights on which the 12 signatory nations agreed.

The Specter of NAFTA

In the United States, the negotiations over the TPP were seen by many to be haunted by the ghost of the North American Free Trade Agreement (NAFTA). The 1993 NAFTA agreement opened a debate in the United States that has never ended; namely, whether that agreement benefitted the U.S. supporters and opponents of NAFTA have very different answers to that question, which may result from the fact that they understand the question differently. Supporters of trade liberalization stress that NAFTA has benefitted both the American and the Mexican economies. Opponents emphasize that the United States lost jobs to Mexico and that the loss of good-paying, middle-class manufacturing jobs has served as a significant restraint on overall wage rates, especially for the lower 40% of the labor force. In 1994, Ross Perot coined an especially memorable phrase—"that giant sucking sound from the south"—to describe the job losses in the United States that would ensue if NAFTA were approved. In contrast, President Clinton promised that NAFTA would create jobs in the United States. NAFTA did create jobs, but overall the number of jobs lost outweighed the number of new jobs.[5] Most obvious to average workers were jobs lost as American companies set up plants in Mexico.

The Impact on Jobs of Trade Liberalization

NAFTA has come to represent something much more than a free trade agreement. Rather, it has become shorthand for the impact of international trade (Scott 2013). The reaction in some quarters against liberalizing international trade stems from the perception that average working people face job losses and gain little as gains accrue elsewhere (Keller and Utar 2016). Job losses are not only visible to the average person but are real although until recently not widely acknowledged. In looking at the impact of trade with China on the United States from 1990–2007, researchers at MIT found that "rising imports cause higher unemployment, lower labor force participation, and reduced wages in local labor markets that house import competing manufacturing industries" (Autor 2013). This study found that import shocks triggered a decline in wages outside of the manufacturing sector, presumably due to the downward pressure on wages in other sectors from the job losses in manufacturing. The authors noted that reductions "in both employment and wage levels lead to a steep drop in the average earnings of households" and concluded that these changes "contributed to rising transfer payments through multiple federal and state programs, revealing an important margin of adjustment to trade that the literature has largely overlooked" (Autor 2013: 2159).

The major difference between 1993 and 2015 is that today the public is aware that there are winners and losers from increasing international trade. Rather than confront that reality, many proponents of liberalizing trade often give reasons to explain job losses that are unpersuasive, most notably that technology is the reason for these job losses. Within any one country, new technology may result in job losses, but technology is not the reason for job losses in the higher-wage country when a company opens up a production facility (using the same level of machinery and technology) in a lower wage country; for instance, when shoe and garment manufacturers shifted production to low-wage countries in Asia. The attraction was extremely low wages. If proponents of further liberalizing trade often refuse to confront the reality of job losses, opponents exhibit a similar blindness. FTAs per se are not the reason for job losses. The demise of the shoe and garment industries in the United States was not due to an FTA.[6] Rather, it was simply that because it was so much cheaper to produce these labor-intensive products in low-wage Asian countries, it made economic sense for companies to shift production abroad, even in the absence of an FTA and with tariff barriers still in effect.

The Rights in Labor Clauses

The furor during the negotiations over NAFTA about the potential for jobs to move to Mexico because of low labor standards produced an innovation. Decrying low wages and unsafe working conditions, opponents of NAFTA called for fair trade, not simply free trade (Compa 2001). Congress's political response was to include for the first time, a statement in the FTA whereby the three signatory nations agreed to uphold a set of labor standards. This was clearly an afterthought, for this commitment was not expressed in the main body of the agreement. Formally named the North American Agreement on Labor Cooperation (NAALC), it is often (and aptly) labeled the NAFTA "side agreement." The idea of inserting a clause relating to labor in an FTA led logically to the question of what standards or principles should be included. In 1993, there was no obviously correct answer. One could argue in favor of listing general principles, or of referring to specific ILO conventions, or listing standards with reference to national law and practice (but that would then have led to the question of which country's law and practice should govern). Within five years, a ready answer became available.

The device of tying trade privileges to adherence to labor standards failed when proposed in December 1996 at the first World Trade Organization (WTO) Ministerial meeting in Singapore. The position taken by the WTO ministers was that there was a specialized UN body that handled such matters, the International Labour Organization (ILO), and as a result the WTO should not consider them.[7] Despite this setback, the adoption in 1998 at the International Labour Conference of the ILO Declaration on Fundamental Principles and Rights at Work proved over time a to be the vehicle for linking trade to workers' rights. The 1998 ILO Declaration contained a succinct statement of four fundamental principles that had been agreed to and receiving overwhelming tripartite acceptance at the June 1998 International Labour Conference. The 1998 Declaration set out four rights, "the principles concerning the fundamental rights which are the subject of those Conventions," namely:

- freedom of association and the effective recognition of the right to collective bargaining
- the elimination of all forms of forced or compulsory labour
- the effective abolition of child labour
- the elimination of discrimination in respect of employment and occupation

Those drafting an FTA and needing language for a labor clause turned readily to the principles set out in the 1998 ILO Declaration. Similarly, the "Labour" section of the June 2000 UN Global Compact lists four principles and repeats word-for-word the language of the 1998 ILO Declaration. Drafters of labor clauses in FTAs quickly seized upon the non-controversial four fundamental principles found in the 1998 ILO Declaration. All U.S. FTAs after 1998 have utilized them.

More recently, the four fundamental principles in the 1998 ILO Declaration have been cast in another light. In June 2011, the UN Human Rights Council unanimously endorsed the Guiding Principles for the Implementation of the UN "Protect, Respect and Remedy" Framework. This clarifies the meaning of the corporate responsibility to respect human rights. To do so, there had to be clarification of what rights businesses are supposed to respect. The Guiding Principles states that this obligation "refers to internationally recognized human rights" and lists the International Bill of Human Rights and the ILO's 1998 Declaration, and in the commentary expressly states these "coupled with the principles concerning fundamental rights in the eight ILO core conventions as set out in the Declaration of Fundamental Principles and Rights at Work" set the benchmarks for assessing human rights impacts.

From this, it seems evident that the UN Human Rights Council viewed the meaning of the ILO Declaration's four fundamental principles as incorporating the principles flowing from the eight core conventions (Bellace 2014: 184).

Meaning of the ILO Fundamental Principles

The 1998 Declaration expressly links eight core conventions to the four fundamental principles, two for each principle:

- Convention No. 87, Freedom of Association and Protection of the Right to Organise (1948) and Convention No. 98, Right to Organise and Collective Bargaining (1948)
- Convention No. 29, Forced Labour (1930) and Convention No. 105 Abolition of Forced Labour (1957)
- Convention No. 138, Minimum Age (1973) and Convention No. 182, Worst Forms of Child Labour (1999)
- Convention No. 100, Equal Remuneration (1951) and Convention No. 111, Discrimination (Employment and Occupation) (1958)

This linkage, however, is expressed in an elliptical fashion. Since most, but not all ILO member states, had ratified these eight core conventions, the International Labour Conference in 1998 considered the obligations of the non-ratifying states, such as the United States. The Declaration in Section 2 states that "all Members, even if they have not ratified the Conventions in question, have an obligation arising from the very fact of membership" in the ILO "to promote and to realize ….the principles concerning the fundamental rights which are the subject of those Conventions."

This direction given the member states is so elliptical as to raise questions. Regardless of whether they have ratified the core conventions, the member states are to promote and realize principles concerning rights expressed in the core conventions with each linked to one of the four principles. The question that logically arises is whether there is some difference in content or in specific obligations between a principle and the linked core conventions, and if so, what difference. There is no answer to that question. But the fact that the United States has refrained from linking the core conventions to the ILO principles in U.S. FTAs, whereas the European Union has not, indicates that there is an open question on this issue and opinions vary as to the answer (Bellace 2016).

Equally intriguing is the slight variation on the articulation of these four principles in the TPP[8] (Chapter 19.3), where it states:

> Each Party shall adopt and maintain in its statutes and regulations, and practices thereunder, the following rights as stated in the ILO Declaration: (a) freedom of association and the effective recognition of the right to collective bargaining; (b) the elimination of all forms of forced or compulsory labour; (c) the effective abolition of child labour and, for the purposes of this Agreement, a prohibition on the worst forms of child labour; and (d) the elimination of discrimination in respect of employment and occupation.

It is curious that the recital of the four principles is almost word-for-word the same as the recital in the Declaration itself, but with regard to child labor, it adds that there is a prohibition on the worst forms of child labor. Yet in 1999 upon the adoption of ILO Convention No. 182, Worst Forms of Child Labour, that convention was immediately linked to the principle regarding abolition of child labor in the 1998 Declaration (Bellace 2014: 179). For those familiar with the ILO's four fundamental principles, it was clear that the meaning of child labor is set forth in ILO Conventions Nos. 138 and 182. Why then was there a reason to insert the title of Convention No. 182 into the TPP's labor clause? One wonders if this implies that the detailed minimum age convention, No. 138, Minimum Age for Entry into Employment, does not govern the notion of what is child labor. That some country apparently felt a need to specify a particular convention as a way of explaining what one of the four worker rights meant might indicate that some of the signatories did not think that these four rights have specific meanings.

In light of the position taken by the 2011 UN Human Rights Council in its Guidelines on business and human rights, that the four fundamental principles are dependent on the eight core conventions for meaning, the question arises of how the 12 signatory nations view the four worker rights listed in Chapter 19, the labor clause of the TPP, and in particular, whether they are simply general concepts or phrases with specific legal meaning.

The Meaning of the Rights in Labor Clauses

This use of ILO principles in FTA labor clauses at first glance indicates governments' acceptance of these principles. Thus this link between FTA labor clauses and the 1998 ILO Declaration, which expressly refers to eight core conventions, implies a common understanding of what these principles mean. But, as has been noted, "in application, this link runs the risk of implementing inconsistent practices, namely, if the States or bodies established by trade agreements use the ILO instruments, as incorporated in their provisions, under a normative or legal meaning that deviates from that previously provided by the ILO supervisory bodies" (Agusti-Panareda, Ebert, and LeClerq 2014: 6).

Recent events in Cambodia illustrate that this risk is not merely theoretical. In 2001 the United States and Cambodia signed a Bilateral Textile Agreement that set quotas for textile imports into the United States. In extending this agreement in 2002, the U.S. Trade Representative's office stated: "The nine percent increase for 2002 reflects Cambodia's progress towards ensuring that working conditions in its garment sector are in substantial compliance with internationally recognized labor standards and provisions of Cambodia's labor law" and noted that the ILO and the United States had projects underway "assisting Cambodia with the implementation of its labor law."[9] Cambodia has ratified all eight of the ILO's core conventions. Yet in February 2014, at a time of violent labor confrontations in the garment industry, the Cambodian Federation of Employers and Business Associations (CAMFEBA) and the Garment Manufacturers Association in Cambodia (GMAC) ran large advertisements in local media saying the public had been misled over the right to strike. The ads stated: "The right to strike is not provided for in … [the ILO's Convention 87 on Freedom of Association] and was not intended to be. … Is the right to strike therefore a fundamental right? NO. The right to strike is NOT a fundamental right."[10] Called upon to respond, Tim de Meyer, a senior ILO official in Asia stated:

> The claims that the right to strike is not a fundamental right and that C. 87 does not establish a right to strike are not consistent with the position taken by the International Labour Organization and its tripartite constituency as a whole (i.e., governments, employers and workers) over a period of at least the last 60+ years.[11]

Although this statement is factually correct, the CAMFEBA (employers' federation) vice president, Sandra D'Amico, demanded that the ILO retract the statement since the "remarks in *The Phnom Penh Post* do not reflect global developments, tripartite consensus or interpretation of right to strike and convention 87."[12] This pressure from employers in a country dependent on exports (Kolben 2014) led the government to propose curbing workers' freedom of association. This April, the legislature passed a Law on Trade Unions that not only sets numerous conditions on the formation of unions but also severely constrains workers' ability to strike, for instance by required that workers who want to stage a protest receive permission from the factory owner to do so or face arrest (Cheang 2016).

This one example illustrates the problem of using the same language in an FTA's labor clause when the parties do not share a common understanding of a fundamental right. A problem arises concerning compatibility between decentralized systems of labor regulation and the global system based on ILO standards as defined within the ILO. If employers support an FTA with one understanding of what the labor clause means, and workers accept that FTA with another understanding of what the labor clause means, it calls into question whether the labor clause in FTAs has any utility beyond mere political convenience. As FTAs increase, this question of the meaning of the labor clauses in FTAs is likely to become more troubling. For instance, the TPP includes Vietnam, a country that does not permit free trade unions (although because of this glaring inconsistency with the principle of freedom of association, the TPP stipulates a "consistency" plan for Vietnam).

Governments' Commitments Under Labor Clauses

During the negotiating phase of the TPP, it was evident that certain countries fell far short of observing the fundamental principles. To stave off criticism that these countries would be permitted to benefit from this FTA, the United States devised a new approach, labeled a "consistency plan," whereby the United States stipulated certain actions the targeted country must take before the country would benefit from the tariff reductions in the TPP. This approach is unusual for a multi-state FTA in that it appears in a side agreement between the United States and the country concerned as a bilateral commitment. There are three separate consistency plans, tailored to circumstances in the given country; namely, Vietnam, Malaysia, and Brunei.

Such labor clauses assume the signatories' willingness to guarantee the stipulated rights and to enforce them. Moreover, a labor clause assumes that one country is willing and able to monitor labor rights in another country and to enforce the labor provisions of the FTA if labor rights are violated in that other country. Both assumptions can be questioned. In November 2014, in studying FTAS then in effect the U.S. General Accounting Office (U.S. GAO) identified challenges to worker rights and pointedly noted "weaknesses in the Office of the U.S. Trade Representative's and DOL's monitoring and enforcement of FTA labor provisions," which even with some improvements between 2009 and 2014 still lacked a "strategic approach" and were "inconsistent with labor provisions in the FTAs" (U.S. GAO 2014: 1).

There is ample evidence that labor rights continue to be violated in countries that have entered into bilateral trade agreements with the United States. The resources for bolstering domestic enforcement of labor laws, investigating violations of trade pact labor provisions, and adjudicating complaints are often lacking. In some countries, ratifying the ILO's core conventions seems to have been the price to be paid to gain trading rights. Cambodia is one example where preferential textile trading rights was the goal. There is little indication that the Cambodian government in 2001 was truly committed to reforming its industrial relations and labor law system. As discussed above, the current adversarial and sometimes violent labor situation is indicative of continuing labor struggles as workers seek to gain recognition of their rights.

More troubling is the uneven pattern of U.S. activity in supporting labor rights in other countries, in part due to the priorities of different administrations (governments) and in part due to the perception of the severity of the violation.[13] The International Labor Affairs Bureau (ILAB) of the U.S. Department of Labor is tasked with providing technical assistance to other countries to assist them in improving labor standards and to monitor and regulate workplaces. To accomplish this, ILAB must be funded adequately. A cursory glance at the funding of ILAB since 1990, a period when there has been a great increase in the number of countries under FTAs that need to be monitored and assisted, reveals an erratic funding pattern indicative of lack of sustained political commitment on the part of the United States to undertake its responsibilities under the labor clauses of FTAs (Callaghan 2009). The cost of hiring, training, deploying, and retaining staff sufficiently knowledgeable about international labor standards, industrial relations, and conditions in lesser developed countries to work to assist those countries in improving labor standards is substantial. One must question why the government of a country would, on its own, seek to do this, especially when it is duplicating work already being done by the ILO and when easy alternatives, such as contributing to a special ILO fund to permit it to intensify its efforts in certain countries, are available.[14]

The innovation of consistency plans for those countries failing to comply with the four fundamental rights carries its own problems. The consistency plans are bilateral agreements between the United States and an individual country (Brunei, Malaysia, and Vietnam). Even more problematic is the approach itself, whereby one country goes into another country to assist it in applying international labor standards, an approach that is somewhat redolent of a colonial power telling its colony how to have better administration. There is a danger that the more advanced country providing technical assistance will utilize the meanings attached to specific international standards as generally understood in its own country, not necessarily the same as those understood by the global system. For the United States, this becomes particularly problematic with regard to freedom of association where the United States not only has not ratified Conventions Nos. 87 and 98 but where current law falls very short of assuring the rights guaranteed by those conventions. Particularly anomalous is the fact that United States has ratified only two of the eight core conventions, whereas Malaysia and Vietnam each have ratified five of the eight ILO core conventions, and are subject to oversight by the ILO supervisory bodies.

Other issues arise with this approach, such as which international labor standards the richer country focuses on. For instance, the United States has expended significant effort to reduce child labor in the countries with which it trades but had made much less effort to bolster freedom of association and collective bargaining. This may result from the individual rights perspective of the United States, or simply from the realization that encouraging the formation of unions and the development of collective bargaining might be controversial at home. But it is perplexing that a country can proceed to assist another country with implementing international labor standards without fully committing to do so at home.

Monitoring and Enforcement

Without precise, agreed upon definitions of what the standards in the labor clause mean, it is difficult to monitor compliance. Egregious violations will be noted, such as the murder of union officials or the deaths of workers in a factory fire where the fire escapes were locked. But even in these instances, enforcement may prove problematic as the experience of the United States itself demonstrates. Since 1993, no country that has signed an FTA with the United States has ever been fined or had its trade privileges revoked, even when severe infringements of freedom of association have occurred. The complaints procedure is designed to make enforcement difficult. Aggrieved workers must find another government that is willing to put forward their complaint (as their own government is extremely unlikely to do so). For labor complaints, there is nothing equivalent to the Investor-State Dispute Resolution process in the TPP or any FTA. Even if a complaint is filed, it languishes for years as discussions occur. Only one case has even reached the arbitration stage, that of Guatemala, and after eight years a decision is still pending.

The Undercutting of Any Meaning of the Labor Standards in the TPP

There has been little discussion until recently about the meaning of international labor standards. The Cambodia dispute mentioned above highlights one area that has become controversial; namely, freedom of association and the right to strike. The 1998 ILO Declaration uses the term "freedom of association" but does not expressly include the term "right to strike." Those who see a distinction between accepting the principles of the 1998 ILO Declaration and ratifying the linked core conventions view the principles as something more general, less specific than the conventions. At the 2012 International Labour Conference, major controversy arose when the Employers' Group asserted that Convention No. 87 does not protect the right to strike. This controversy is still simmering, although it is widely acknowledged that it is difficult to comprehend how workers in practice can utilize their freedom of association to form unions and achieve effective collective bargaining without the ability to strike. In light of this controversy over the critical question of what ILO fundamental rights mean, it is startling to find buried in the TPP's labor clause, in footnote 3 of Chapter 19, the statement that "The obligations set out in Article 19.3 (Labour Rights), as they relate to the ILO, refer only to the ILO Declaration." Only the cognoscenti would recognize that this means the four principles found in the ILO's 1998 Declaration have been severed from the eight core conventions. Without being moored to specific meanings, terms such as "freedom of association" mean what any signatory nation thinks it means. This then reveals the hidden agenda in the curious re-phrasing of the four rights in TPP's labor clause, Chapter 19. As noted above, in the recital of child labor, there was added the phrase that the worst forms of child labor are not permitted. Since the prohibition of "child labour" in Chapter 19 is now de-linked from the core conventions mentioned in the 1998 Declaration, we only know that children under 18 should not be permitted to engage in work dangerous to their physical or moral being. We do not know at what age they are permitted to work in a factory (which is stipulated in ILO Convention No. 138).

A further dilution of meaningful standards appears in footnote 4 of Chapter 19, which states: "To establish a violation of an obligation under Article 19.3.1 (Labour Rights) or Article 19.3.2, a Party must demonstrate that the other Party has failed to adopt or maintain a statute, regulation or practice in a manner affecting trade or investment between the Parties." It can be argued that trade and investment between the United States and many countries does not relate to compliance with worker rights or even the enforcement of basic safety standards. Bangladesh would be one such example. Another example is Guatemala where, in the current arbitration hearings, the position has been taken that egregious violence, including the murder of union officials, while conceivably dampening workers' enthusiasm for unions, does not affect trade. If this is correct, then footnote 4 in Chapter 19 would seem to make it extremely difficult to establish a violation of rights that have been stated in the main text of Chapter 19.

Even vaguer is the section on safety and health, Chapter 19.3.2 wherein it states: "Each Party shall adopt and maintain statutes and regulations, and practices thereunder, governing acceptable conditions of work with respect to minimum wages, hours of work, and occupational safety and health." That this section might have any useful meaning is undercut by footnote 5 in Chapter 19, which states: "For greater certainty, this obligation relates to the establishment by a Party in its statutes, regulations and practices thereunder, of acceptable conditions of work as determined by that Party." In essence then, "the highest standard labor clause" as proclaimed by the U.S. Trade Representative takes the position that acceptable conditions of work are determined by each country, and there is no obligation to meet any agreed upon minimum. As long as a country has not set higher standards for work conditions, it is difficult to see how any workers in that country who are aggrieved by unsafe and sweatshop conditions of work can complain under the TPP.

Once the text of Chapter 19 is read closely, and in particular, the footnotes are read, one might reasonably conclude that the labor clause in the TPP is essentially meaningless.

Next-Generation Labor Clauses

A truly "high standard" labor clause would specify that signatories are bound to observe, apply and enforce internationally recognized labor standards, and in particular the four fundamental principles set forth in the 1998 ILO Declaration *and* the linked eight core conventions. To avoid confusing and/or conflicting interpretations, such a clause would expressly state that "these rights should be understood in a manner consistent with that expressed by the ILO's supervisory system." Such a statement included in an FTA would lay the basis for a justiciable claim. But more is needed.

Dispute Resolution

Rather than the tortuously slow process for complaints arising under the labor clause of existing FTAs, what is needed is a form of dispute resolution with a long history in American labor relations; namely, arbitration. A speedy dispute resolution process needs to be designed. For instance, parties alleging violations of these rights would go through state-to-state dispute resolution, but only for two years. If at that point the matter was not resolved in a manner satisfactory to the complainants, the complainants (not a state party) could invoke arbitration, with that process to take place within one year. FTAs already include Investor-State Dispute Resolution Mechanisms for non-labor matters that permit foreign corporations to sue governments for what a company sees as unfair treatment. Rather than going to court, investors can compel governments to litigate the matter in a private arbitration system. This mechanism could be adapted to apply to labor matters. Workers in a country alleging that they are affected by unfair competition because of non-compliance with worker rights or labor standards in another signatory country could invoke a Workers–State Dispute Resolution mechanism and take their complaint to a private arbitration panel. Besides speed, another advantage would be that arbitrators would be selected for their knowledge of law (including international labor law) and practice, experience with labor matters, neutrality, and lack of susceptibility to political influence, factors that are not always found among judges in many countries, including the United States (Compa 2016; Kolben 2007: 244).

Whether such provisions would actually result in arbitration hearings is not clear, but the distinct possibility of going to arbitration would most likely have a beneficial effect in moving the offending state to take action (Banks 2011: 95). These provisions would persuade parties to observe labor standards that comport with the meaning generally accorded them in international law, and would likely compel parties to resolve violations promptly thus realizing workers' rights. The TPP as it currently stands is unlikely to produce that result.

Endnotes

[1] The 12 countries currently participating in the TPP are Australia, Brunei, Canada, Chile, Japan, Malaysia, Mexico, New Zealand, Peru, Singapore, the United States, and Vietnam.

[2] The TPP can come into force if at least six of the original signatories who between them account for 85% of the group's GDP ratify the agreement. That can only occur if both the United States and Japan ratify the TPP since together they account for 80% of the group's GDP.

[3] Canada, Mexico, Singapore, Australia, Peru, and Chile.

[4] This phrase has been used consistently by the office of the U.S. Trade Representative on its website. See, e.g., ustr.gov/sites/default/files/TPP-Strategic-Importance-of-TPP-Fact-Sheet.pdf

[5] Economists disagree as to the exact number, as it is very difficult to disentangle the fact of the free trade agreement from other factors that would have resulted in manufacturing moving to Mexico, such as the devaluation of the peso.

[6] An example is Bangladesh. The Office of the U.S. Trade Representative reported that in 2013 Bangladesh imported $712 million in goods from the United States whereas it exported $5.4 billion in goods to the United States (ustr.gov/countries-regions/south-central-asia/bangladesh). The United States has no FTA with Bangladesh.

[7] Singapore Ministerial Declaration Adopted on 13 December 1996 WT/MIN(96)/DEC, available at www.wto.org/english/thewto_e/minist_e/min96_e/wtodec_e.htm

[8] ustr.gov/trade-agreements/free-trade-agreements/trans-pacific-partnership/tpp-full-text

[9] January 7, 2002, press release available at fordschool.umich.edu/rsie/acit/LaborStandards/LaborInUSCambodiaTextile.pdf

[10] Shane Worrell, "Groups tell ILO to retract 'right to strike' claim." The *Phnom Penh Post*, Feb 6, 2014. www.phnompenhpost.com/national/groups-tell-ilo-retract-%E2%80%98right-strike%E2%80%99-claim

[11] Id.

[12] Id.

[13] The author believes that a majority of people view child labor as particularly abhorrent but the suppression of unions, and even the murder of trade union officers, is not deemed as serious.

[14] The International Programme for the Elimination of Child Labour is one example where several advanced countries did contribute voluntarily to a fund so that the ILO could expand its efforts to eradicate child labor.

References

Agusti-Panareda, Jordi, Franz Christian Ebert, and Desirée LeClerq. 2014. *Labour Provisions in Free Trade Agreements: Fostering Their Consistency with the ILO Standards System.* International Labour Office, March 2014. www.ilo.org/wcmsp5/groups/public/---dgreports/---inst/documents/genericdocument/%20wcms_237940.pdf%20

Agusti-Panareda, Jordi, Franz Christian Ebert, and Desirée LeClerq. 2015. "ILO Labor Standards and Trade Agreements: A Case for Consistency," *Comparative Labor Law and Policy Journal*, Vol. 36, no. 3 (Spring 2015), pp. 347-380.

Autor, David H., David Dorn, and Gordon H. Hanson. 2013. "The China Syndrome: Local Labor Market Effects of Import Competition in the United States," *American Economic Review*, Vol. 103 no. 6 (Oct 2013), pgs. 2121-2168.

Banks, Kevin. 2011. Trade, Labor and International Governance: An Inquiry into the Potential Effectiveness of the New International Labor Law, *Berkeley Journal of Employment and Labor Law,* Vo. 32, no. , pp. 45-142. scholarship.law.berkeley.edu/cgi/viewcontent.cgi?article=1432&context=bjell

Bellace, Janice R. 2014. "Human Rights at Work: The Need for Definitional Coherence in the Global Governance System," *International Journal of Comparative Labour Law and Industrial Relations,* Vol. 30, no. 2, pp. 175-198.

Bellace, Janice R. 2016. "Back to the Future: Freedom of Association, the Right to Strike and National Law," *King's Law Journal,* Vol. 27, no. 1 pp. 24-45.

Cheang, Sopheng. 2016. "Cambodia lawmakers pass proposed law making rules for unions," *The Seattle Times,* April 4, 2016. www.seattletimes.com/business/cambodian-parliament-passes-disputed-trade-union-law

Compa, Lance A. 2001. "Free trade, fair trade, and the battle for labor rights," in Lowell Turner, Harry Katz, and Ronald Hurd eds., *Rekindling the Movement: Labor's Quest for Relevance in the 21st Century,* Ithaca, NY 314-338. ILR Press, pp. 314-338.

Compa, Lance A. 2016. "How to Make the Trans-Pacific Partnership Work for Workers and Communities," *The Nation*, January 14, 2016. www.thenation.com/article/how-to-make-the-trans-pacific-partnership-work-for-workers-and-communities

Council on Foreign Relations. 2016. "*The Future of U.S. Trade and the Trans-Pacific Partnership: A Conversation with Michael Froman: TPP and American Leadership in the Pacific.*" June 20, 2016. www.cfr.org/trade/future-us-trade-trans-pacific-partnership-conversation-michael-froman/p37973

Keller, Wolfgang, and Hâle Utar. 2016. *International Trade and Job Polarization: Evidence at the Worker-Level*, National Bureau of Economic Research, Working Paper 22315. www.nber.org/papers/w22315

Kolben, Kevin. 2007. "Integrative Linkage: Combining Public and Private Regulatory Approaches in the Design of Trade and Labor Regimes," *Harvard International Law Journal*, Vol. 48, no. 1 (Winter 2007) pp. 203-256.

Kolben, Kevin. 2014. "Trade, Monitoring and the ILO: Working to Improve Conditions in Cambodia's Garment Factories." *Yale Human Rights and Development Journal*, Vol. 7, no. 1 (2014) pp. 79-107. digitalcommons.law.yale.edu/cgi/viewcontent.cgi?article=1043&context=yhrdlj

Scott, Robert E. 2013. *NAFTA's Legacy: Growing U.S. Trade Deficits Cost 682,900 Jobs*, Economic Policy Institute (Dec. 17, 2013). www.epi.org/publication/nafta-legacy-growing-us-trade-deficits-cost-682900-jobs

U.S. Government Accountability Office (U.S. GAO). *Free Trade Agreements: U.S. Partners Are Addressing Labor Commitments, But More Monitoring and Enforcement Are Needed.* GAO-15-160, November 2014. www.gao.gov/assets/670/666787.pdf

U.S. Trade Representative. 2016. *The Trans-Pacific Partnership.* ustr.gov/trade-agreements/free-trade-agreements/trans-pacific-partnership/tpp-full-text

VIII. $15 Minimum Wages: What Can the Research Tell Us?

Nonprofit Government-Funded Human Services and the 2015–2016 New York Minimum Wage Campaigns

James A. Parrott
Roosevelt House Public Policy Institute at Hunter College

New York State enacted a statewide $15 minimum wage in April 2016. Among low-wage sectors heavily affected is the social assistance, or human services, sector with approximately 315,000 workers statewide. Most of these workers are employed by nonprofits providing services under government contract. As minimum wage floors rise, service delivery costs for labor-intensive nonprofits can rise significantly. This paper examines the efforts in 2015–2016 of nonprofit human service providers to get New York City and New York State to increase contract funding in $5 billion in human services contracts for the purpose of funding higher mandated wages.

Wage Floors and the New York City and State Nonprofit Human Services Sector

In the spring of 2013, New York State enacted an increase in its minimum wage to $9.00 an hour in three annual steps, with the first step an increase from the federal $7.25 minimum wage to $8.00 an hour on the last day of 2013 and the third step reaching $9.00 on the last day of 2015. In the 2015 legislative session, bills were introduced, but not enacted, to authorize the City of New York and other local governments to set their own minimum wages, up to a level 25% or 30% above the statewide minimum.

New York City Mayor Bill de Blasio took office in January 2014, and in September 2014 issued an executive order to expand and increase the minimum wage level to $13.13 an hour for the city's living wage law that applied to companies receiving substantial economic development benefits from the city. He did not act at that time to adjust the city's living wage law applicable to selected city service contracts. The service contract living wage affected workers in seven occupational categories (childcare, Head Start, temporary clerical, food service, building services, home health care, and workers providing services to people with cerebral palsy), with wage levels ranging from $10.00 an hour plus a $1.50 supplement for benefits if employer health insurance is not provided (for childcare, Head Start, and home care workers; and workers providing services to people with cerebral palsy) to $26.20 plus $10.46 for supplemental benefits for the highest class of building cleaner.[1]

With the exception of workers in these occupational areas, employees of nonprofit organizations providing human services under contract to the City of New York are exempt from coverage under the city's service contract living wage. An effort to include such workers in 2002 was rebuffed by nonprofit leaders on the grounds that the city under Mayor Michael Bloomberg was not prepared to adjust contract funding so that nonprofits could pay the higher wages.[2]

In late 2014, leading New York City nonprofit human services organizations began to organize to press Mayor de Blasio to increase human services contract funding connected to a first-ever wage floor in this sector.[3] Because of the importance of additional funding, the new campaign focused on the city budget rather than amending the

---

Author's address: 47-49 East 65th Street, New York, NY 10065 (At the time of this writing, Parrott was with the Fiscal Policy Institute.)

service contract living wage. By early 2015, Mayor de Blasio was advocating a phased-in $15 minimum wage for New York City. In his FY 2016 budget proposal released in early May 2015, he included funding for an $11.50 wage floor for human services contracts plus a 2.5% cost-of-living adjustment (COLA) for human services contract workers paid above the $11.50 wage floor level. This was the first wage increase or COLA for such contracts since the onset of the 2008–2009 Great Recession. Dating back at least to the early 2000s (and possibly further back), the city had periodically provided additional funding to nonprofit human services contractors to give employee wage increases following the wage pattern negotiated under municipal labor contracts. Under de Blasio's predecessor, Mayor Bloomberg, the municipal labor contract round covering the years 2008–2010 had not been completed by the time Bloomberg left office at the end of 2013.

Establishing an $11.50 wage floor for human services contract workers occurred at the same time the city modified its collective bargaining agreement with District Council 37 of AFSCME to raise the pay of its lowest-paid members—school crossing guards—to $11.50 an hour. The city's FY 2016 budget, adopted by the city council in late June 2015, included $25 million to fund the $11.50 human services contract wage floor, effective July 1, 2015, and $29 million for the 2.5% COLA. In addition, at the request of the advocacy campaign, the new budget included $5 million to develop a sector-wide education and training system to provide career ladder services and supports for 80,000 nonprofit contract workers employed on city contracts.

Influenced by the national Fight for $15 campaign and the political clout of two large affiliates of SEIU, Governor Andrew Cuomo used his executive authority to convene a wage board in May 2015 to investigate the adequacy of compensation in New York's rapidly growing fast-food industry. The wage board held four hearings around the state in June and took testimony from policy experts and scores of fast-food workers. In July, the wage board recommended a phased-in $15 wage floor to be reached at the end of 2018 in New York City and by July 2021 in the rest of the state. The final wage order was issued in September 2015 and covered an estimated 122,000 workers in fast-food chain restaurants.[4]

At the time the final fast-food wage board recommendation was issued, the governor first suggested legislation to enact an across-the-board phased-in $15 minimum wage increase for all workers on the same timetable as specified in the fast-food wage order. The legislation would be introduced in January for the 2016 state legislative session. Using his executive authority, the governor announced in November that he was implementing a $15 minimum wage for all workers in the State University of New York (SUNY) system.

Not to be outdone in the jostling for policy leadership in proposing higher wages, Mayor de Blasio announced in early January 2016 that all city employees (mainly affecting a few thousand school crossing guards and seasonal parks employees) and all city-funded nonprofit human services contract workers would be covered by a $15 minimum wage. This announcement benefited approximately 50,000 workers—20,000 direct city employees and 30,000 full-time equivalent positions in the contracted human services sector. The mayor proposed the same phase-in schedule as for fast-food workers in New York City—hence, reaching $15 by December 31, 2018, and proposed that the city fund the wage increase for nonprofit contract workers. Mayor de Blasio included this funding—which would rise to $100 million annually when $15 was fully phased in—in his preliminary FY 2017 budget in January 2016 and in his executive budget proposal released in late April 2016.

The FY 2017 budget was adopted by the city council in mid-June. As was the case with the funding in the FY 2016 budget, the council supported the mayor's proposal to increase contract funding for the purpose of phasing in a $15 wage floor for nonprofit human services workers.

The nonprofit advocates pressing for city funding for higher wages launched a parallel statewide campaign in the fall of 2015 to seek increased state government wage increase-related funding in state and local government human services contracts. The coalition leaders released a public policy advocacy report outlining the need in December 2015.[5] The report noted that because nonprofits do not have any pricing power, an unfunded mandate to raise the wage floor affecting many of their workers might lead to increased caseloads and the curtailment of some services—and could lead to the closure of some nonprofits. The report also noted that if nonprofit human services workers were excluded from coverage under the higher wage floor, recruitment and retention issues that already pose a challenge to the sector would only be exacerbated. Based on preliminary estimates of $1.5 billion in state human services contracts, the coalition report estimated that a fully phased-in $15 minimum wage would require approximately $250 to$300 million in additional funding, including some provision for moderate wage increases for workers paid slightly above $15 an hour, to avoid undue wage compression.[6]

As at the New York City level, part of the organizing work of the nonprofit coalition at the state level involved convincing other nonprofits that had previously opposed minimum wage increases to support the legislation for a $15 minimum wage and to press for the additional state funding to fund higher wages. Some of the organizations from the sector providing services to the developmentally disabled—a sector largely funded by Medicaid reimbursements but with average wages in the $11 to $12 range—had testified against a higher minimum wage for fast-food workers on the grounds that they could lose workers to the fast-food sector because fast-food work involves considerably less responsibility and is less stressful than serving the developmentally disabled population. For several years, the developmentally disabled sector had sought greater state Medicaid funding without success. During the early 2016 campaign, the sector serving the developmentally disabled supported the higher state minimum wage.

Several discussions were held with representatives of the state Budget Division and the governor's office to make the case for additional contract and Medicaid funding. One of the complicating factors was a rigid 2% state operating budget spending cap that the governor had instituted early in his first term. In practice, the 2% overall spending cap meant that funding for human services usually had declined in nominal terms from the prior year because Medicaid and local school aid—two budget categories accounting for about half of all state operating expenditures—had higher growth limits that had worked out to range from a little under 4% to 6%. This meant that human services and other spending (e.g., general-purpose local aid, higher education funding, and funding for parks and transportation services) had been falling in real terms in most years since 2012.

Throughout the campaign, the nonprofit coalition held fast in support of the minimum wage increase and actively worked to secure its passage even in the absence of any firm commitment from the governor's office to increase nonprofit funding.

The legislature, particularly the Democratic-controlled Assembly, was much more sympathetic and included in its budget proposal $200 million for additional funding for human services contracts and other state-supported mental hygiene and home health care services. While the Republican-controlled Senate publicly resisted the governor's $15 minimum wage proposal, there was some support from key Republicans for additional human services contract funding should the minimum wage increase be approved. However, the leadership of neither chamber was ready during the 2016 budget session to seriously challenge the governor on his 2% spending limit.

In the broader New York $15 minimum wage campaign in the early months of 2016, the governor and advocates highlighted the need for a higher minimum wage to enable workers to meet family budget needs and stressed the positive economic impact of additional spending by the 3 million-plus workers who would benefit from a phased-in wage increase.[7] To counter arguments asserting a negative impact on businesses, the Institute for Research on Labor and Employment (at UC Berkeley) released a report presenting an integrated analysis of the combined effects on business operations and consumer spending that showed small net employment effects and a significant improvement in living standards for low-wage workers.[8]

Even though Governor Cuomo's legislative proposal included a slower phase-in schedule for the suburban areas outside of New York City and for upstate, opponents argued that upstate businesses would have a much harder time adjusting to higher wages given weaker economic conditions than in the downstate region. Lower overall median wage levels for upstate areas were cited. In response, the Fiscal Policy Institute pointed out in a report that at a detailed occupational level for the largest low-wage occupations (e.g., retail clerks, cashiers, or stock clerks), median wage levels in upstate metropolitan areas were often within 5% to 10% of corresponding wage levels in New York City and the downstate suburbs and that in some jobs, such as retail sales, the highest median wage levels were in upstate areas.[9]

As a condition of supporting the higher state minimum wage, the Senate Republicans insisted on a longer phase-in period for the upstate region north and west of Westchester County. The enacted legislation calls for a $15 minimum wage to be reached in New York City at the end of 2018, in the three largest suburban counties (Nassau, Suffolk, and Westchester) by the end of 2021, and a level of $12.50 in the rest of the state by the end of 2020 with subsequent annual adjustments based on a yet-to-be-determined formula decided by the state budget office and labor department with the increases continuing until $15 is reached.[10]

The enacted New York legislation did not include provision for any inflation adjustment once the $15 level is reached. Without a built-in inflation adjustment factor, the purchasing power of the eventual $15 wage floor will be eroded in future years unless there are periodic adjustments.

The FY 2017 state budget that was adopted as part of a broader agreement to raise the minimum wage included $19 million to adjust Medicaid reimbursement rates for workers providing Medicaid-funded home health care and developmental disability services and to fund minimum wage increases for certain school districts and providers of special education services. The budget cost of these increases will rise to $588 million by state FY 2020.[11] However, the budget did not include funding to increase amounts for state human services contracts whose employees will be covered by the statewide minimum wage increases.

The reason for this disparate treatment of private sector workers providing state-funded human and health services likely has something to do with the fact that Local 1199 of SEIU represents tens of thousands of affected home health care and nursing home workers. The union also invested heavily in the pro-minimum wage campaign and was a co-leader along with 32BJ SEIU (the large New York City-based building service workers local) in the progressive campaign in support of the increase. Most of the nonprofit human services workers not covered by any spending increase in the budget are not unionized.

The Cuomo administration's position with regard to human services contracts is that they will consider possible contract funding adjustments on a case-by-case basis in instances where there is financial "hardship" for the organization. While the governor has voiced concern that some nonprofit executives are "highly paid," suggesting there is no need in such cases to increase government funding to raise pay for low-paid workers, some state officials indicate that the state's 2% spending cap is a key factor in complicating the ability of the state to fund human services contract increases.[12]

Purchase of Service Contracting

Settlement houses and various religious and other charitable organizations have a long history in New York City of providing a range of human services to low-income communities. The provision of government-funded human services grew sharply in the wake of the Great Society anti-poverty programs in the 1960s, and the scale of service provision has expanded significantly in the decades since, for a variety of reasons, and to meet a wide range of needs. Sometimes this programmatic growth was in response to developments such as increased female labor force participation, the AIDS epidemic, the increased number of elderly, or the deinstitutionalization of those with mental health issues. To take another example, federal welfare reform in 1996 pushed many mothers of young children into the paid workforce, increasing the need for childcare subsidies and after-school programs, and the TANF block grant that was part of welfare reform provided a funding source for more services for low-income households.

Almost from the start of the expansion in the 1960s, publicly funded human services provision has been channeled through nonprofit organizations rather than directly by employees of state or local government agencies. There was already a sizable infrastructure of charitable service providers, and new community-based organizations emerged as government funding increased. The decision to fund services through nonprofits was done to build on existing service networks, to foster community-oriented approaches to service delivery, and partly to keep costs down and limit the further growth in government agencies that had expanded in response to pressures from the civil rights movement and urban unrest in the 1960s. Nonprofits expanded their capacity along with growing human services needs and government funding opportunities. Nearly a quarter century ago, Smith and Lipsky noted that nonprofits had become the "favored tool of public service delivery."[13]

Beginning in the 1970s, and accelerating in the 1980s, the federal government devolved greater responsibility for providing and funding human services to states, which in turn relied on nonprofit contractors to deliver services.[14] As Fabricant and Fisher have written, while federal social policies in the 1960s and early1970s relied on contracting to overcome the limits of private charity and local and state governments, by the Reagan years and into George H.W. Bush's term, federal policies reoriented the federal role in order to end social welfare expansion and reduce reliance on unionized public sector workers.[15]

Within New York, the state has tended to rely heavily on local governments, particularly New York City, to share funding responsibility for public assistance programs and the delivery of most human services. In recent years, New York City has contracted out over $4 billion in human services delivery annually, primarily to nonprofit organizations. An estimated 80,000 full- and part-time workers are supported under city contracts.[16] The state funds a significant share of contract funding in certain areas, like foster care, family homeless shelters, domestic violence, and AIDS-related services, but many program areas are largely, if not entirely, city-funded. In addition, the state directly contracts with New York City nonprofits to provide about $1 billion in services.[17]

Most nonprofit organizations providing human services have become heavily dependent on government contracts. While some service providers have become less connected to the neighborhoods from which they evolved years ago, many continue to reflect community concerns but have been limited by inadequate or inflexible public funding. For the past quarter century, many nonprofits have relied on government purchase of service contracts for 80% or more of their funding. For example, in 1993, 80% of the combined budgets of the 38-member settlement houses within the United Neighborhood Houses umbrella organization were from government contracts, and 80% of that contract total represented contracts with the City of New York.[18]

Most New York human services nonprofits were experiencing serious financial challenges even before the Great Recession of 2008–2009. The recession sharply reduced state and local tax revenues that resulted in several years of budget cuts falling heavily on the human services contract sector. With the historically weak economic and employment recovery of 2009 to 2013, human services funding was reduced at a time when economic hardships remained elevated, as indicated by the two-thirds increase in supplemental nutrition assistance recipiency in New York State during the recession and recovery.[19] In New York City, while overall city-funded expenditures rose by a slight 2% on an inflation-adjusted basis from 2008 to 2013, real spending on human services fell by 8%.[20]

Over the years, state and city contracts generally have left New York nonprofit human services providers in a precarious financial position. Following the abrupt bankruptcy and dissolution of one of the largest citywide multi-service nonprofits—the $250-million-a-year FEGS organization—a task force established by the umbrella Human Services Council reported that government contracts cover, on average, only about 80% of each dollar of "true program delivery costs." City contracts, for example, typically provide limited funding for indirect costs. A major nonprofit providing criminal justice services notes that while the federal government provides a 19% indirect rate, many city contracts allow only 10% or even as little as 5% for indirect costs.[21]

Moreover, both the city and the state routinely make contract payments several months following service delivery, forcing nonprofits to undertake costly borrowing that is not reimbursable.[22]

A comprehensive financial analysis of New York's nonprofit sector found that government accounted for 90% of total revenues for the median (by revenue) health and human services nonprofit. The financial analysis also determined that 18% of the city's nonprofit health and human services providers were insolvent, and 50% had almost no cash reserves, allowing little margin for error. The Oliver Wyman/SeaChange Capital Partners report concluded, "government contracts ... virtually guarantee a deficit. Government contracts also create working capital needs because funding arrives after expenses are paid. These funds are also subject to unpredictable delays."[23]

The Nonprofit Human Services Workforce—Educated, Dedicated, Discriminated, and Underpaid

There are about 315,000 social assistance (also called human services in this paper) workers in New York State, the great majority of whom work for nonprofit organizations.[24] Human services employment grew by 150,000, or 91%, from 1990 to 2015, with 100,000 of that job gain occurring between 1990 and 2004, with an increase of 50,000 jobs since 2004. While New York City accounts for 57% (179,000) of statewide human services employment (vs. its roughly 45% share of all nonagricultural jobs), job growth in this sector was 116% outside of New York City from 1990 to 2015 compared to a 76% increase in New York City.[25]

FIGURE 1
New York State and New York City Social Assistance Employment 1990–2015

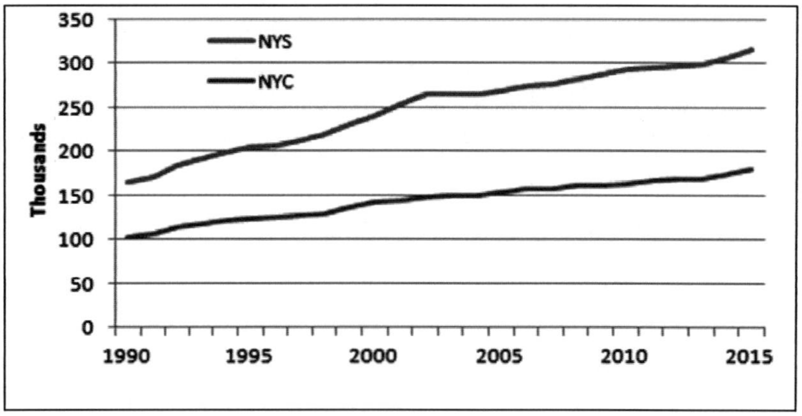

Average annual wages in the social assistance or human services sector were about $27,100 in 2015, lower than in all other New York industries except for restaurants ($26,400) and grocery stores ($25,000). About half of the workers were paid less than $15 an hour in 2014, and 30% earned less than $10.50 an hour.[26] Teachers and support workers in childcare centers are particularly low paid. Occupational employment statistics data indicate that half of all preschool teachers were paid less than $13.80 an hour in 2014, and 90% of childcare workers made less than $14.50 an hour.

Even workers in the human services sector who are not at the bottom of the sector's pay scale are significantly underpaid considering their education, skill levels, and the responsibilities of their jobs. Most social workers and mental health, substance abuse, and other counselors in the field have a master's degree and need to meet and maintain professional certifications or licensing requirements, yet are paid well under $50,000 annually. Workers with comparable education and responsibilities but employed directly by government or elsewhere in the private sector frequently have salaries well above that level.[27]

Two-thirds of human services workers have some level of college education, with 45% holding four-year bachelor's degrees or higher. Women make up 82% of the statewide workforce. People of color account for 50% of human services workers in the state and an even larger portion in New York City, where they account for 75% of the workforce.[28]

According to an analysis by the UC Berkeley Center for Labor Research and Education, more than three out of five New York human services workers paid below $15 an hour receive some form of public assistance, or a family member does.[29]

Compensation is so low for New York's human services workers for many reasons, chief among them historical gender- and race-based pay discrimination, general lack of unionization, and the weak bargaining position of their nonprofit employers vis-à-vis state and local government contracting agencies.

Not surprisingly, given the low pay and demanding jobs within the nonprofit human services sector, many nonprofits struggle to maintain a highly skilled and dedicated workforce. Recruitment and retention is a serious challenge for many organizations. Based on a survey of its member organizations providing child welfare and childcare services around the state, the Council of Family and Child Caring Agencies reported an average annual turnover of over 30% in 2014.[30]

Next Steps for the New York City Career Ladder Campaign and the Statewide $15 and Funding Campaign

Earlier in this paper, a brief overview was provided of the New York City and statewide campaigns of human services organizations pressing for a higher wage floor and increased government funding to enable nonprofits to pay the higher wages. This section will provide further information on those campaigns as well as an assessment of the efficacy of the advocacy efforts and suggestions for further work.

The city campaign met with much greater success partly because of supportive elected officials, particularly Mayor de Blasio and the speaker of the city council, Melissa Mark-Viverito. Still, the campaign's goal of funding higher wages for city service contract workers was not an initiative proposed by any elected official; it was advanced by a newly formed coalition of human services organizations and advocates. One of the coalition's biggest challenges was organizing the sector itself, a sector populated by several large multi-service and often citywide agencies as well as many mid-size and smaller community-based nonprofits. The sector had never before come together to address the low pay of its workforce, and its budget-related advocacy agenda had usually consisted of demands for a general COLA.

The campaign initiated by the Federation of Protestant Welfare Agencies and the Fiscal Policy Institute was called the Career Ladder Project and had the twin objectives of raising the wage floor in stages to $15 an hour, with the city providing the requisite funding to pay higher wages; and establishing a sector-wide career ladder system. The focus was clearly on raising the pay and enhancing career opportunities for the lower-paid half of the sector's workforce, which is overwhelmingly comprised of people of color, mostly women, who live in some of the city's poorer neighborhoods. In that regard, the purpose was squarely in line with the mayor's emphasis on raising wages and living standards for the city's least advantaged. Higher wages for all workers in the lower and middle tiers of the sector would be necessary to complement an effective career ladder system in order to reward workers increasing their skills and education. Connecting the funding requirement to the wage floor increase was critical in getting major nonprofits on board with this agenda, which represented a significant change in direction from previous advocacy efforts.

The Human Services Council, which for several years had led nonprofit campaigns seeking COLAs, again sought a COLA as the 2015 budget season unfolded, but soon began working with the Career Ladder Project (CLP). By the time the mayor's executive budget proposal was finalized, the two campaigns came together behind a three-part initiative: a funded $11.50 wage floor (to be effective with the start of the next city fiscal year, July 1, 2015), a 2.5% COLA for all other workers, and $5 million to fund the development and start-up of a human services sector-wide education and training system to provide career ladder opportunities. The three components were incorporated as proposed in the adopted FY 2016 budget. Since then, HSC, FPWA, and FPI have combined their efforts to continue to broaden coalition-building efforts at both the city and state levels.

As the human services coalition moved to the state level in the fall of 2015, it dealt with a similar challenge in enlisting the cooperation of a diverse set of nonprofit service providers to actively support the governor's $15 statewide minimum wage proposal, again by linking it to the need for increased government funding. The statewide campaign went by the hashtag "#15 and Funding," so named to connect to the visibility and organizing success of the broader Fight for $15 campaign and to stress the need for increased contract funding.

Despite the governor's enthusiastic support for raising the minimum wage, his conservative fiscal policy posed a huge obstacle to securing the required state funding to enable nonprofit service contractors to pay higher wages. While the Assembly majority and many senators from both parties supported increased funding related to the wage increase, funding was not part of the compromise agreement because it was not a priority for the governor. In the end, the coalition just did not have sufficient support among key Senate Republicans to force the governor's hand.

The waning days of March 2016 witnessed the spectacle of New York and California racing to be the first state to enact a $15 minimum wage. In the end, it appeared to be a photo-finish tie. However, when it comes to ensuring that no workers are left behind, California clearly came out ahead when its agreement included a commitment to provide budget funding so that state-supported childcare workers, as well as workers providing Medicaid-reimbursed services, would see the higher wage floor. [31]

The New York State coalition has its work cut out for it in more effectively pressing the case for increased human services funding in the year ahead. Some of the coalition partners are also preparing to work with a broader set of advocates, such as leading community organizing groups, public sector unions, and others concerned about the limits on New York's state and local government spending in pushing to raise or end the governor's self-imposed 2% spending cap. The human services coalition is also starting to frame its advocacy in terms of the need for pay equity and state government responsibility for the publicly funded human services workforce. Now that the across-the-board $15 minimum wage floor has been established, the coalition will focus directly on funding concerns. A more direct focus on funding is likely to garner broader and deeper support among nonprofit service providers.[32]

Because of the cumulative effects of persistent under-funding, there is also a critical need not only to increase contract funding to enable nonprofits to pay higher wages up to the new wage floor but also to raise the pay of human services workers now paid above $15 an hour but underpaid given their education, skills, and job responsibilities. In New York, this has often been referred to as the wage compression issue, referring to the need to increase pay for workers above the minimum to avoid a morale-challenging compression of wage differentials. The wage compression issue is another manifestation of a pay equity problem because many better-educated and female-dominated counselor and social work positions fall into this category.

An additional dimension that the coalition needs to effectively address involves local government-funded human services contracts elsewhere in the state outside of New York City. In the state of New York, county governments constitute social service districts and fund contracts for child welfare, housing, after-school care, childcare, and youth and senior services. These services are mainly provided by nonprofits facing the same funding constraints as state- or New York City-funded nonprofits. Because the state's local property tax cap (set at the lesser of inflation or 2%) severely constrains county government finances, the coalition is planning to advocate for increased state aid to county governments for the purposes of increasing amounts in human services contracts to nonprofits facing higher wage costs. This should help develop allies among county executives and local legislators and provide a more visible local constituency in the districts of Senate Republicans.

Because childcare is such a large and important service area and because early childhood educators and childcare workers are so poorly paid, there is also a compelling need to develop a childcare sector-specific policy agenda as the higher wage floor is phased in. One of the most pressing policy priorities likely will be increased funding for childcare subsidies, both to address the long-standing need to increase the number of families who can receive subsidies but also to increase the amount of the subsidy to reflect the higher costs associated with the minimum wage increase. A related priority will be to increase the state's child and dependent care tax credit to assist families of modest means who pay for childcare out of their own pockets as childcare fees rise.

As part of the advocacy to secure additional state funding, the coalition would be well-served by being able to provide reasonable estimates of potential cost savings that will accrue to state and local governments in the form of reduced public assistance spending as higher wages lift workers above eligibility thresholds. Such estimates should also reflect the higher personal income (or lower state EITC credits) and sales tax revenues that would be generated as the wage floor rises.

Finally, at the New York City level, FPWA and FPI are continuing to work with the city to develop the sector-wide career ladder system that will include not only education and training programs but also career and financial aid counseling and childcare support services for the children of workers attending classes. Yet to be addressed is whether it makes sense to advocate for some form of career advancement supports for state-funded human services workers.

Government Responsibility for the Human Services Contract Workforce and for Fully Funding Service Contracts

State government relies heavily on nonprofits to provide a broad range of important public services. Because its contracting practices and funding commitments have contributed to the substandard compensation levels for the nonprofit workforce and to the precarious financial condition of most nonprofit service contractors, state government should acknowledge its role as the indirect employer and its responsibility to fully fund the provision of contracted public services.

Governor Cuomo has talked about the need to recruit and invest in the next generation of state government workers, but he needs to recognize that the state also has an obligation to change its funding and contracting practices so that nonprofits can better recruit, retain, and develop a skilled workforce providing services that are just as important a state responsibility as providing education, transportation, or public safety services. Fully funding the cost of contracted public services will also likely result in higher-quality and more effective and efficient services.

The state also has an obligation to assist county governments in raising the pay of county-funded service contract workers and in fully funding such contracts.

The City of New York has accepted some responsibility for its contracted workforce in budgeting in its four-year financial plan for covering the full cost of a phased-in $15 wage floor. However, the city still needs to make a commitment to address the compression issue by increasing the pay of demonstrably underpaid workers that now receive more than $15 an hour.

There are many other aspects of purchase-of-service contracting that the city and the state need to address, including higher indirect rates and timelier contract and payment processing and the need for more expense flexibility in performance-based contracts.

Endnotes

[1] The city's Service Contract Living Wage Law is Section 6-109 of the NYC Administrative Code.

[2] Several nonprofit leaders later regretted that decision because it contributed to a perception that nonprofit leaders were not willing to fight for better wages for their employees.

[3] This campaign was initially co-led by the Federation of Protestant Welfare Agencies (FPWA), an umbrella group, and the Fiscal Policy Institute (FPI), a nonprofit New York State and local fiscal and economic policy research and education organization. By the spring of 2015, the leadership group also included the umbrella Human Services Council (HSC).

[4] See *Report of the Fast Food Wage Board to the NYS Commissioner of Labor*, July 2015, labor.ny.gov/workerprotection/laborstandards/pdfs/Fast-Food-Wage-Board-Report.pdf

[5] FPWA, Human Services Council and Fiscal Policy Institute, *A Fair Wage for Human Services Workers: Ensuring a Government-Funded $15 per Hour Minimum Wage for Human Services Workers Throughout New York State*, December 2015.

[6] Ibid.

[7] David Cooper, *Raising the New York State Minimum Wage to $15 by July 2021 Would Lift Wages for 3.2 Million Workers*, Economic Policy Institute, EPI Briefing Paper #416, January 5, 2016; National Employment Law Project, *How Much Do New York's Workers Need? At Least $15 per Hour—Both Upstate and Down*, Fact Sheet, January 2016.

[8] Michael Reich, Sylvia Allegretto, Ken Jacobs, and Claire Montialoux, *The Effects of a $15 Minimum Wage in New York State*, UC Berkeley Institute for Research on Labor and Employment, Policy Brief, March 2016.

[9] FPI, *Upstate–Downstate Wage Differentials Are Relatively Small in Low-Wage Occupations*, March 24, 2016.

[10] For a chart with the timing and amounts of increases in the wage floor for fast-food and all workers in New York State, see Patrick McGeehan, "New York's Path to $15 Minimum Wage: Uneven, and Bumpy," *New York Times*, April 1, 2016. For a discussion of the reduced economic impact of the slower minimum wage phase-in for the upstate regions, see Edward Krudy, "New York's Two-Tier System Offers Testing Ground in Minimum Wage Debate," Reuters, June 30, 2016.

[11] Office of the New York State Comptroller, *Report on the State Fiscal Year 2016–17 Enacted Budget Financial Plan and Capital Program and Financing Plan*, July 2016, Appendix A.

[12] The governor has never raised the issue of executive compensation at hospitals even though the state's voluntary hospitals receive billions of dollars in Medicaid and other state funding. With state tax revenues growing 4% to 5% annually while state operating spending is subject to a 2% cap, lack of revenues is certainly not a constraint. The governor has supported a series of tax cuts as part of his austere approach to state spending. From state FY 2011 to 2017, general fund expenditures for human services contracts grew by only 0.3% annually. After inflation, this represents a 1% annual decline. In contrast, during the recession and its immediate aftermath,

inflation-adjusted state general fund spending on human services contracts increased by 3.1% annually, according to the Fiscal Policy Institute.

[13] Steven Rathgeb Smith and Michael Lipsky, *Nonprofits for Hire. The Welfare State in the Age of Contracting*, Cambridge, MA: Harvard University Press, 1993, p. 11.

[14] Michael B. Fabricant and Robert Fisher, *Settlement Houses Under Siege. The Struggle to Sustain Community Organizations in New York City*, New York: Columbia University Press, 2002, pp. 76-77.

[15] Ibid, pp. 77-78.

[16] FPI estimate.

[17] FPI analysis of contract data from the city's Office of Management and Budget, the New York City comptroller, and the New York State comptroller.

[18] Fabricant and Fisher, pp. 78-81.

[19] Fiscal Policy Institute, *A Shared Opportunity Agenda, New York State Economic and Fiscal Outlook, 2015–2016*, January 2015, pp. 38-41.

[20] James Parrott, *Beyond Balance: Forward-Looking Budget Priorities for New York City*, in *Toward a 21st Century City for All: Progressive Policies for New York City in 2013 and Beyond*, edited by John H. Mollenkopf, New York: The Center for Urban Research, The Graduate Center, City University of New York, 2013, pp. 75-76.

[21] Joanne Page, "The Trouble with City Contracts," New York Nonprofit Media, May 13, 2016.

[22] Human Services Council, *New York Nonprofits in the Aftermath of FEGS: A Call to Action*, February 2016.

[23] Oliver Wyman and SeaChange Capital Partners, *Risk Management for Nonprofits*, March 2016.

[24] The 2012 Economic Census indicates that 84% of the employment in the New York City social assistance sector is in nonprofit organizations. For 2014, Quarterly Census of Employment and Wage data indicate that 72% of New York City social assistance employment is in nonprofits.

[25] New York State, Current Employment Statistics.

[26] Annual wage data from QCEW; hourly wage data from FPI analysis of 2010–2014 Current Population Survey.

[27] FPWA, Human Services Council, and Fiscal Policy Institute, *A Fair Wage for Human Services Workers: Ensuring a Government-Funded $15 per Hour Minimum Wage for Human Services Workers Throughout New York State*, December 2015.

[28] FPI analysis of 2010–2014 Current Population Survey Outgoing Rotation Group data provided by the Economic Policy Institute.

[29] Ken Jacobs, Ian Perry, and Jenifer MacGillvary, *The Public Cost of Low Wages in New York*, UC Berkeley Center for Labor Research and Education, January 2016.

[30] Mary Jane Dessables and James F. Purcell, *Workforce Compensation Report*, Council of Family and Child Caring Agencies, April 2014.

[31] The enacted 2016–17 California state budget includes a total of $210 million to cover the first step of the increased minimum wage (rising from $10 to $10.50 on January 1, 2017) and to provide an increase for direct-contracted and voucher-based childcare providers. See the enacted budget summary (ebudget.ca.gov/2016-17/Enacted/BudgetSummary/BSS/BSS.html) and enacted budget detail (ebudget.ca.gov/2016-17/Enacted/agencies.html) for K–12 education and health and human services.

[32] Governor Andrew Cuomo has often championed women's issues, so an aggressive pay equity campaign may meet with greater success. In addition to pushing through a $15 minimum wage increase in the 2016 session, the governor actively campaigned to make New York the fifth state in the nation to provide family leave insurance. [33]

[33] See the public policy recommendations in the Human Services Council Issue Brief, *Recommitting to the Nonprofit Sector: Creating a Path to Sustainability Through Policy Change*, Fall 2013.

IX. Work Challenges in First Nations Communities

The Occupational Structure of the American Indian/Alaska Native Workforce

Jacob Wise
University of Minnesota

Dick Todd
Federal Reserve Bank of Minneapolis

Carolyn A. Liebler
University of Minnesota

We analyze the occupational structure of the non-Hispanic American Indian/Alaska Native (AIAN) workforce in the United States, relative to the non-Hispanic White workforce, using Public Use Census Microdata. AIAN workers are generally over-represented in low-skilled occupations and under-represented in high-skilled occupations, relative to White workers. This pattern is stronger among men than among women and stronger among single-race AIANs than multiple-race AIANs. AIAN occupational dissimilarity has not declined substantially since 1980. Controlling for individual differences in factors such as education, age, location, and language proficiency accounts for a significant proportion of AIAN under-representation in high-education occupations.

Introduction

Occupational structure is a useful social indicator. Group differences in occupational attainment may signal inefficiencies that significantly reduce economic productivity, such as labor market discrimination or suboptimal investment in education. Occupational differences can also mediate other adverse social and economic disadvantages—occupations differ in average pay, sensitivity to business cycles, health risks, prestige, status, and authority.

We analyze the occupational structure of the non-Hispanic American Indian/Alaska Native (AIAN) workforce in the United States. Although racial and ethnic differences in occupational patterns have been documented and analyzed for decades (e.g., Blau and Duncan 1967), few studies have focused on the occupational structure of the AIAN workforce, and none that we know of have separately examined both AIAN workers who identify as single-race and AIAN workers who identify as multiple-race.

A detailed analysis of AIAN occupational structure is timely in light of economic and social changes that have affected the AIAN workforce in recent decades. Economies in many reservations and homeland areas have grown rapidly (albeit from a low base) in recent decades (Akee and Taylor 2014). This directly affects many AIANs—about one-fifth of AIAN individuals (single-race and multiple-race combined) lived on a reservation or other homeland as

Author's address: Dick Todd, Federal Reserve Bank of Minneapolis, 90 Hennepin Avenue, Minneapolis, MN 55401 of 2010 (Norris, Vines, and Hoeffel 2012). Since the 1970s, tribal colleges have expanded significantly (see www.collegefund.org/ content/tcu_timeline), and there has been a general increase in AIAN educational attainment (see Figure 5 later in this paper). In the broader economy, the occupational distribution of the general workforce has changed significantly in response to deindustrialization and rising service employment.

Measurement changes have also added to the value of an update on occupation and race. Partly as a result of the shift in the general occupational distribution, the Standard Occupational Classification system used by federal agencies was developed in 1977 and updated as of 1980, 2000, and 2010 (Emmel and Cosca 2010). In 1997, the federal government broadened the definition of AIAN to include Central and South American indigenous people and required that multiple-race responses be allowed (Office of Management and Budget 1997). In the censuses of 2000 and 2010, individuals were instructed to "mark one or more" races. In the 2010 Census, there were about 2.3 million individuals who identified as AIAN in combination with another race or races, as well as 2.9 million who identified as AIAN alone (Norris, Vines, and Hoeffel 2012).

In this paper, we address three research questions about non-Hispanic AIAN occupational stratification.[1] First is the occupational distribution of AIAN workers different than that of Whites, now and since 1980? We show that it is and that AIAN workers share many occupational patterns long observed among other racial or ethnic minorities. We find that the pattern of occupational dissimilarity between AIAN workers and White workers is stronger among men than among women (although still significant among women). We do not find that AIAN occupational dissimilarity has declined substantially since 1980, though results about changes over time are relatively tenuous due to changes in measurement and racial identification (see Liebler, Bhaskar, and Porter 2016).

Second, in which occupations are AIAN workers under-represented relative to White workers? In which are they over-represented? We compare single-race Whites to single-race and multiple-race AIANs. Using Census 2000 and the 2008–2012 American Community Survey (ACS), we find that AIAN workers are generally over-represented in low-skilled occupations and under-represented in high-skilled occupations, relative to White workers. This distinction is less pronounced for multiple-race AIANs than for single-race AIANs.

Third, do standard demographic factors account for the under-representation of AIAN workers in high-education occupations (relative to White workers)? Among the observable factors that may account for AIAN–White differences (including age, location, and language proficiency), we find that gaps in educational attainment are the most important. Controlling for individual differences in these factors reduces the degree of AIAN under-representation but fails to account for it fully. We regard the remaining occupational structure differences we find between AIAN and White workers as a sign that deeper social and economic issues may continue to restrain the well-being of the AIAN population.

Previous Studies

In their landmark study *The American Occupational Structure*, Blau and Duncan (1967) documented basic occupational differences between Whites and non-Whites (94% of whom were "Negro" in their sample (p. 207)). After ranking 17 occupations primarily by the median income and education of incumbents in 1962 (p. 26), they found that the occupation status typical for non-Whites was not only different from that of Whites but also "far inferior to that of whites" (p. 209). Although lower educational attainment explained part of this difference, it remained large "even when the lower social origin, education, and first occupation of Negroes [had] been taken into account" (p. 209).

Blau and Duncan's key themes have been confirmed in subsequent studies. Different occupational patterns for minority workers, as opposed to majority workers, have repeatedly been found, with minority workers generally holding lower status or lower paid occupations.

However, few papers study the AIAN workforce. An exception is work by Alonso-Villar, Del Rio, and Gradin (ADG) (2012), who included "Native Americans" (non-Hispanic single-race American Indians, Alaska Natives, Hawaiians, or Pacific Islanders) among six racial/ethnic groups and included all non-Hispanic multiracial individuals as "other." They found substantial occupational dissimilarity between Native Americans and the population, with "Native Americans ... concentrated in lower-paid occupations" (p. 190).

Our research is similar to ADG (2012) but is different in at least five ways that allow us to build on their results. First, we focus specifically on the AIAN workforce, as opposed to a composite group of "Native Americans." Second, we implement recently developed statistical tests (Allen, Burgess, Davidson, and Windmeijer 2015) to assess the significance of the differences in dissimilarity we report. Third, because occupational patterns remain quite different by sex, we compare within sexes rather than comparing to the overall workforce of both sexes, as in ADG. Fourth, we examine not only the single-race AIAN workforce but also the multiple-race AIAN workforce. ADG study only single-race AIANs, which introduces bias because single-race AIANs are not representative of the entire AIAN group (Liebler and Halpern-Manners 2008). Finally, when modeling factors associated with occupational differences, we use an education-based ranking of occupations as the dependent variable (rather than regional differences), so that our regressions directly shed light on factors related to the tendency for AIAN workers to be concentrated in low-skill sectors.

Data

We focus our analyses on the American Community Survey five-year pooled sample from 2008–2012, which hereafter we will refer to as 2010, its middle year. For a few analyses, we also use additional public-use data sets collected by the Census Bureau: decennial census data from 1980, 1990, and 2000 (5% samples). We accessed all data through the IPUMS USA project at the University of Minnesota (Ruggles et al. 2015). We used weights (PERWT in IPUMS) to create statistics that are nationally representative of persons (Lumley 2004).

Our ability to detect changes over time in occupational dissimilarity is limited by changes in categorizations over time. The race categorization system has changed substantially. Since 2000, the Census has invited respondents to report multiple races and included multiple-race responses in the data, allowing us to begin tracking the multiple-race responses separately from single-race responses. There is evidence suggesting that the AIAN category does not include a consistent set of individuals across the decades (Liebler, Bhaskar, and Porter 2016; Liebler and Ortyl 2014; Passel 1997). In addition, the categorization of occupations was fundamentally changed between the 1990 Census and the 2000 Census and was modified again by 2010 (Norris, Vines, and Hoeffel 2012). Although we have used a constructed variable (OCC2010 in IPUMS) that attempts to map the earlier categories into the contemporary categories, changes in definitions, as well as the evolving nature of jobs in the economy, make perfect mapping impossible.

Throughout the paper, we include all workers age 16 and over. Because young workers have often not completed their education, we have checked that the conclusions based on the regression results we report below are robust to limiting our sample to workers age 25 and up. We find that results for the older workers are consistent with those reported here.

Our focus in this research is on two categories of American Indians/Alaska Natives: people who reported being non-Hispanic American Indian/Alaska Native alone (AIANa), and people who reported being non-Hispanic American Indian/Alaska Native in combination with one or more other races (AIANc). We show results for Hispanic AIANs only in Table 1 and the appendix table; elsewhere we combine Hispanic AIANs with other Hispanics. Hispanic AIAN workers are an especially small group. Also, people who report as AIAN and Hispanic rarely give the same race response in another census, but they often consistently report Hispanic origins (Liebler, Bhaskar, and Porter 2016).

Results

Is the AIAN Occupational Distribution Different?

In Table 1 we show a breakdown of the U.S. labor force in 2000 and 2010 by race, where each category (except where explicitly listed) is single-race and non-Hispanic. From the results in Table 1, we see that there are relatively few AIAN workers—AIAN single-race and multiple-race individuals together comprised 1.43% of the (age 16+) *labor force* but 1.63% of the population in 2010 (authors' calculations). In most of our analyses, we compare AIANa and AIANc individuals to the largest race group in the workforce: single-race Whites.

TABLE 1
Percentage Breakdown of the (Age 16+)
Labor Force by Race, Over Time

	Race	2000	2010
1	White	72.71	65.99
2	African American	10.57	11.45
3	Hispanic	10.48	14.95
4	Asian/PI	3.75	5.14
5	Remainder	1.09	1.04
6	AIANa	0.66	0.58
7	AIANc	0.55	0.58
8	Hispanic AIAN	0.19	0.27
	Total	100.00	100.00

Our first research question is "Is the occupational distribution of AIAN workers different than that of single-race White workers, now and since 1980?" We begin to address this question using Figure 1, in which we plot the distribution of the workforce in 2010 across 26 occupation groups, separating out the results by sex and for non-Hispanic AIAN alone, AIAN in combination, and White.[2]

Some general patterns are evident. Females and males are very differently distributed across occupations, and differences by sex are generally large relative to differences by race. AIANc workers have an occupational distribution that is generally between that of AIANa workers and Whites; the share for AIANc lies between the shares for AIANa and White in 18 of the 26 career categories for men and 17 of 26 for women. Also noteworthy is a tendency toward under-representation of AIANa and AIANc workers of both sexes in traditional "white-collar" occupation categories, such as management, financial specialists, and legal professions, and their over-representation in traditional "blue/pink-collar" fields such as construction, healthcare support, and building/grounds cleaning and maintenance. With respect to our first research question—whether the occupational distribution of AIAN workers is different from White workers—Figure 1 (see next page) presents a mixed picture of gross similarity overall (for each sex) but also many differences occupation-by-occupation.

Nonetheless, the answer to our first question is clear: the AIAN occupational distribution was significantly different from the White occupational distribution in 2010 and each of the three preceding decades. To show this, we use data across all the occupations in Figure 1 to calculate an overall index of occupational dissimilarity between each AIAN group and the corresponding group of single-race White workers. Specifically, we calculate the widely used Duncan index D, defined as follows: For n occupations, we compute the statistic

$$D = (1/2) \sum_{i=1}^{n} |A_i/A - B_i/B|$$

where A (or B) is the total number of individuals of type A (or type B) and A_i (or B_i) is the number of Type A (or B) individuals in occupation i. This widely used index goes back at least to Blau and Duncan (1967), but it has properties that can be undesirable (Watts 1998). As a check on the robustness of our results, we also compute the alternative indices I_p (proposed by Karmel and MacLachlan 1988) and A (proposed by Charles and Grusky 1995). In the text, we note when our results are sensitive to the choice of index.

FIGURE 1
Percentage of Workers of That Race Group Who Worked in Each Type of Occupation in 2010, for Non-Hispanic Single-Race AIANs (AIANa), Multiple-Race AIANs (AIANc), and Whites, Separated by Sex

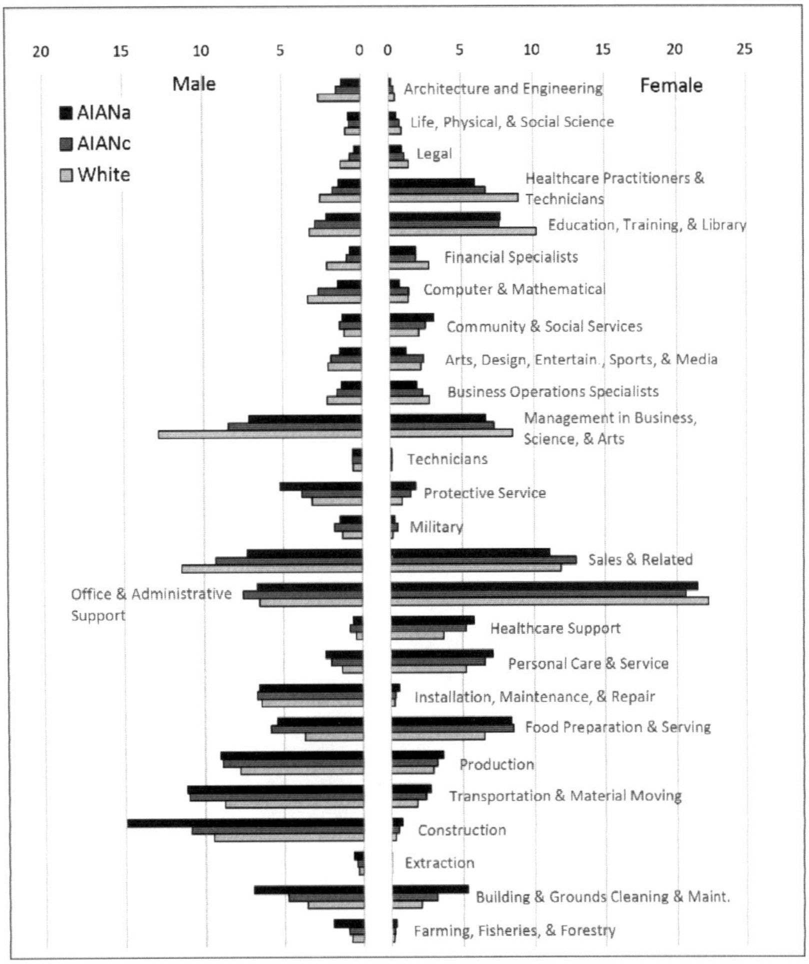

This index of occupational dissimilarity can be interpreted as the percentage of workers who would need to change careers in order to make the AIAN and White occupational distributions identical. In 2010, D is about 16.5 for AIANa workers and 9.9 for AIANc workers. In other words, 16.5% of AIANa workers would need to change occupation categories in order to have the same occupational distribution as White workers. Both of these values are very significantly different from zero ($p < 0.001$ for both), according to the likelihood ratio test described by Allen and colleagues (Allen et al. 2015).

Table 2 shows this index of dissimilarity for AIAN and other racial/ethnic groups over four decades, for males and females combined, as compared to non-Hispanic Whites. All of the index values in the table are significantly different from zero. As in 2010, dissimilarity to single-race Whites in 2000 is smaller for AIANc workers than for AIANa workers. For both 2000 and 2010, the degree of dissimilarity for AIANa workers is closer to that of African American or Asian/Pacific Islander workers than to the value for AIANc workers and is about halfway between the values of AIANc workers and Hispanic workers.

TABLE 2
Index of Dissimilarity (and Standard Errors) for 26 Occupation Categories, Split by Race/Hispanic Origin and Decade [comparisons are to single-race non-Hispanic Whites for whom the index dissimilarity (from themselves) is automatically zero]

	AIANa*	AIANc*	Asian/PI	African American	Hispanic**	Remainder
1980	17.79 (0.37)		16.55 (0.21)	21.62 (0.08)	20.50 (0.10)	9.98 (2.00)
1990	18.16 (0.32)		15.97 (0.14)	20.35 (0.06)	22.43 (0.09)	14.91 (2.03)
2000	16.54 (0.27)	10.15 (0.47)	17.78 (0.12)	19.09 (0.08)	23.30 (0.07)	10.07 (0.31)
2010	16.47 (0.38)	9.92 (0.55)	18.13 (0.11)	19.00 (0.07)	24.14 (0.07)	10.33 (0.35)

*For 1980 and 1990, we report the AIAN data under the AIANa column (the Census did not allow multiple race responses until 2000).

**All Hispanics are grouped together.

Our results show that the degree of AIAN occupational dissimilarity changed little between 2000 and 2010 and we see no clear AIAN trend overall since 1980. This is in contrast to an apparent small but steady decrease in occupational dissimilarity from Whites for African Americans and a steady increase in occupational dissimilarity for Hispanics.

However, intertemporal comparisons should be made with caution because of cross-time changes in measures and race group composition (for example, see Liebler et al. 2014).

We also calculate the AIANa-White dissimilarity index separately for men and women in 2010 because men and women choose different occupations (as highlighted in Figure 1). Similar to the findings reported by Taylor (1994) for the distribution of indigenous Australian workers, we find a lower occupational dissimilarity index between AIANa women and White women (14.5% would need to change occupation groups to achieve parity) than between AIANa men and White men (19.8%), and this difference is statistically significant. However, for women as well as men, the answer to our first question is the same—AIAN workers have a different occupational distribution than single-race White workers.

Dissimilarity indices for AIANa and AIANc workers appear to vary substantially by region within the U.S. In Figure 2 (see next page), we show the occupational index of dissimilarity for AIANa people in 13 regions (defined and discussed by Eschbach 1992). The AIANa occupational dissimilarity index is higher in areas with relatively high AIAN populations than in areas with relatively fewer AIANs. For AIANa workers, the Southwest and North Carolina stand out as having the highest degree of occupational dissimilarity with Whites in the same region, with Alaska, California, and the Basin-Mountain, Northern Plains, and Great Lakes regions also showing high levels of

AIANa-White dissimilarity. For AIANc workers, different areas with high AIAN populations stand out as regions of higher occupational dissimilarity from local Whites—Alaska and the Northern Plains.

There were very disparate results for AIANa versus AIANc workers in the Southwest and North Carolina (which we find to be statistically significant, in tests not shown). In the South, the dissimilarity from local Whites is relatively low for both AIAN groups, and, in Alaska, the dissimilarity is relatively high for both. In the Northern Plains, dissimilarity appears relatively high for AIANc workers relative to Whites (higher than for AIANa workers in five regions and nearly on par with AIANa dissimilarity nationally), but the dissimilarity for AIANa workers there is higher yet. However, neither of these apparent results for the Northern Plains is statistically significant, due to few observations and thus large standard errors. (Standard errors are included in parentheses in the figure.)

FIGURE 2

Occupational Dissimilarity Indices by Region (Blue = AIANa and Green = AIANc)

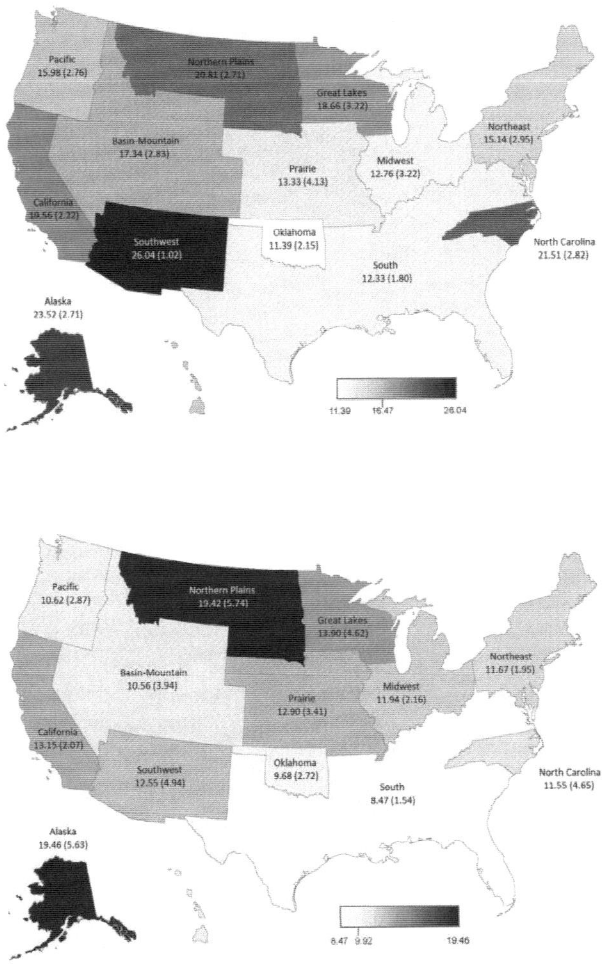

Over- and Under-Representation in Occupations

Our second research question is "In which occupations are AIAN workers under-represented relative to White workers? In which are they over-represented?" To begin answering this question, we return to Figure 1 (previously shown). The occupational categories there are ordered by the fraction of incumbents who had completed at least one year of college, based on the data from 2010. For example, 92.9% of members of the architecture and engineering profession attended college. This was the highest rate of college attendance by labor-force participants in any of the occupation groups, so it is shown at the top. Those in the farming, fisheries, and forestry category, shown at the bottom, had the lowest percentage of incumbents who attended college (16.9%). See Table 4 (later in this paper) for details.

With this ordering, Figure 1 suggests an occupational gap based on education. To investigate the statistical significance of the differences in occupational participation, we display in Figure 3 an index based on the ratio of the AIAN-alone employment proportion to the employment proportion of the White workforce (for men and women combined).

FIGURE 3
Under-/Over-Representation of AIAN Alone, 2010 (calculations and style based on Fox 2003)

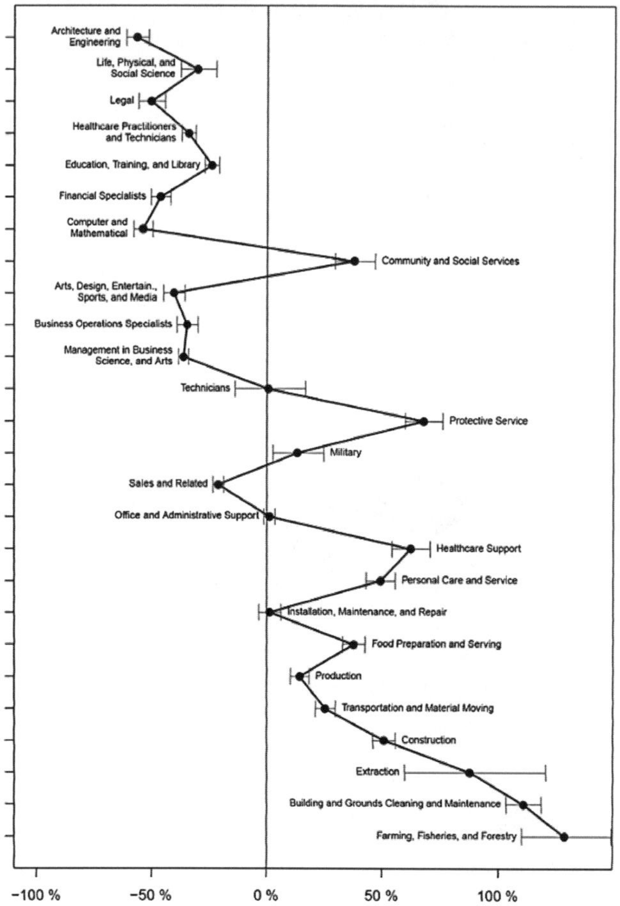

Specifically, for AIANa workers our index for any single occupation takes the value:

$$\frac{\text{share of AIAna workers in the occupation}}{\text{share of White workers in the occupation}} - 1$$

In Figure 4, we also show a parallel index for AIANc workers. Figure 3 includes thin red lines showing the 95% confidence interval for each career category and maintains the education-based ordering of the careers. In the careers in the bottom half, where individuals typically have less education, there is generally over-representation of AIAN workers. This tendency disappears in the mid-level career categories (education-wise) and transitions toward the top of the figure to under-representation in fields where higher levels of education are common.

The ratios in Figure 3 display a distinct "tilt" in the occupational representation of AIANa workers. In nine of the ten lowest categories on the education scale, there is a statistically significant overrepresentation of AIANa workers. In fields like building and grounds cleaning there are twice as many AIANa workers employed, relative to the proportion of Whites in that sector. In the most highly educated occupation categories, AIANa individuals are under-represented in ten of the top eleven (the exception being community and social services). For legal professions in particular, there are 50% fewer AIANa workers employed than one would expect if AIANa participation were proportional to participation by single-race Whites.

FIGURE 4
Under-/Over-Representation of AIAN Over Time
[filled circles represent statistically significant (non-zero) values]

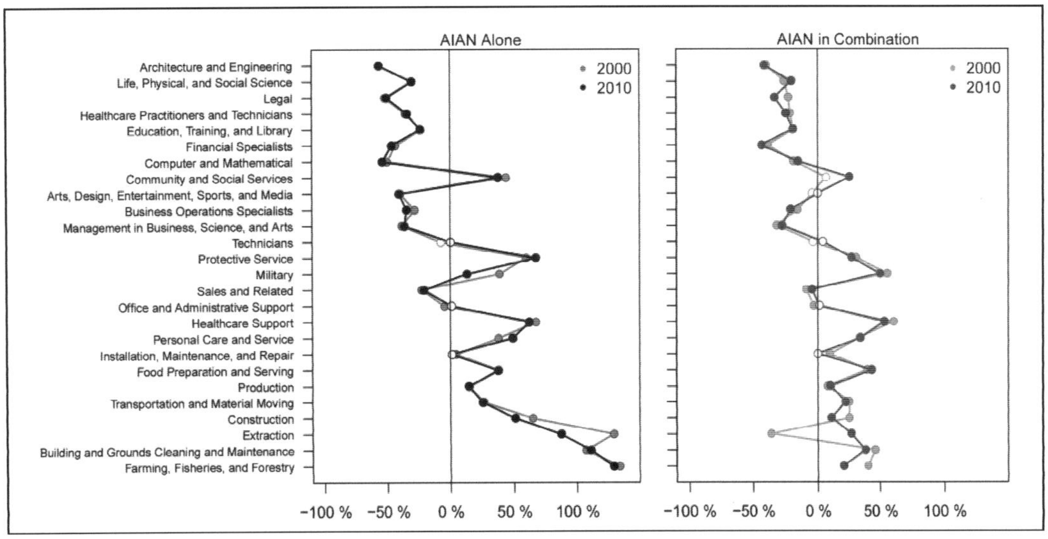

In a version of Figure 3 (available on request) with men and women analyzed separately, the basic pattern is the same, but a few subtleties emerge. For example, AIAN underrepresentation in the legal professions is larger for men than for women. Also, although AIAN workers overall are overrepresented in protective services, this is even more true for AIAN females relative to White females than it is for AIAN males relative to White males.

In Figure 4, we expand Figure 3 to include AIANc workers and to include data for both 2000 and 2010. The results for AIANc individuals show a pattern of statistically significant, education-based occupational disparity that is qualitatively similar to the pattern for AIANa individuals (as seen by the visual "tilt" of both panels). However, the pattern is quantitatively milder for AIANc workers than AIANa workers. In both panels of Figure 4, the observed changes in career categories between 2000 and 2010 are small (with the largest differences, like those in "extraction," mainly due to small cell counts).

Adjusting the Dissimilarity Index for Educational Attainment

To further explore the relationship between education and occupation, we calculated the AIAN-White indices of dissimilarity within each of five education categories: Less than high school, high school degree, some college or associate's degree, bachelor's degree, and more than a bachelor's degree. We show these results for 2010 in Table 3. For example, 12.23% of single-race AIAN workers in the lowest education category would need to change fields (to another of the 26 fields) in order for their occupational distribution to match that of the least-educated White workers. We again find that all occupational dissimilarity values for both AIAN groups in Table 3 represent a statistically significant dissimilarity from Whites ($p < 0.001$).

TABLE 3
Occupational Dissimilarity Index
(and Standard Errors) by Race and Education in 2010

	AIANa	AIANc	White
No HS Deg.	12.23 (1.69)	7.86 (2.44)	0
HS Grad	12.95 (0.74)	7.69 (1.15)	0
Some College	10.97 (1.17)	6.88 (1.32)	0
BA	13.70 (2.05)	10.87 (1.85)	0
BA+	9.79 (2.81)	8.48 (2.27)	0
All	16.47 (0.38)	9.92 (0.55)	0

From the calculated statistics shown in Table 3, we notice that racial comparisons restricted to like-educated workforce members often produce a smaller index of occupational dissimilarity than for the general workforce (shown in the last row). This indicates that differences in educational attainment partly explain the high overall occupational dissimilarity between the AIAN workforce and the White workforce. However, the index of dissimilarity is still quite high within education categories, especially among individuals with a bachelor's degree but no further education.[3] Also note that the results for AIAN in combination again lie between the results for Whites and AIAN alone for each educational category.[4]

A Formal Test to Answer Question 2

Figures 3 and 4 already show a statistically significant pattern of AIAN over-representation in low-education occupations and under-representation in high-education occupations. To provide a clear test of this overall tendency, we construct a binomial regression model, predicting the probability that a given individual is employed in a "highly educated field" (the binomial "success") or not. In defining "highly educated fields," we sort military workers (an industry) back into their original occupation groups. We code "high" education fields as "Architecture and Engineering" through "Management in Business, Science, and Arts" and "low" education fields as "Technicians" through "Farming, Fishing, and Forestry" (Table 4, next page). This dichotomy roughly corresponds to careers with a higher fraction of college-educated participants than in the general workforce. In Table 4, we also show the occupational income (average income of incumbents) of each of the 26 broad occupation groups. Ranking occupations by income instead of education would result in a generally similar definition of high-ranked versus low-ranked occupations.

TABLE 4
Occupational Education and Occupational Income for Occupation Groups, 2008–2012 ACS

Occupation Group	Occupational Education		Occupational Income	
Architecture and Engineering	high	93%	high	$85,155
Life, Physical, and Social Science	high	92%	high	$68,781
Legal	high	92%	high	$115,592
Healthcare Practitioners and Technicians	high	90%	high	$75,637
Education, Training, and Library	high	90%	low	$43,963
Financial Specialists	high	90%	high	$76,898
Computer and Mathematical	high	90%	high	$76,918
Community and Social Services	high	88%	low	$43,022
Arts, Design, Entertainment, Sports, and Media	high	81%	low	$49,298
Business Operations Specialists	high	80%	high	$68,255
Management in Business, Science, and Arts	high	76%	high	$87,336
Technicians	low	63%	low	$52,601
Protective Service	low	61%	low	$48,859
Military	low	59%	low	$44,328
Sales and Related	low	55%	low	$43,321
Office and Administrative Support	low	54%	low	$32,855
Healthcare Support	low	48%	low	$24,817
Personal Care and Service	low	45%	low	$20,723
Installation, Maintenance, and Repair	low	38%	low	$43,477
Food Preparation and Serving	low	33%	low	$15,747
Production	low	28%	low	$35,266
Transportation and Material Moving	low	28%	low	$33,464
Construction	low	26%	low	$36,669
Extraction	low	22%	low	$52,030
Building and Grounds Cleaning and Maintenance	low	22%	low	$22,281
Farming, Fisheries, and Forestry	low	17%	low	$22,249

In Table 5 on the following page, we show the results of a basic regression predicting whether a worker is in a high education occupation based only on their race response. The coefficient estimate for the intercept (-0.502) implies that a White worker in the year 2010 had a

$$\frac{e^{-0.502}}{1+e^{-0.502}} = 37.71\%$$

chance of being employed in a highly educated field. The coefficient for AIANc worker implies a change in the log-odds of being in a highly educated field relative to the odds for the comparison group (Whites), calculated as

$$e^{-0.396} - 1 = -32.70\%$$

Thus, an AIANc worker in the labor force has a probability of 28.95% of being employed in a highly educated field. The results also imply that AIANa individuals in the labor force have an even lower probability, just 24.47%, of being employed in a highly educated field.

TABLE 5
Binomial Regression Predicting Employment in a "Highly Educated Field," for 2010

	Estimate	SE	z value	Pr(>|z|)	
(Intercept)	−0.502	0.001	−565.0	<2e-16	***
AIANc	−0.36	0.010	−38.87	<2e-16	***
AIANa	−0.625	0.011	−58.29	<2e-16	***

Notably, the Wald tests (comparing each coefficient to zero, for which the p-values included refer) and the relative size of the standard errors in Table 5 provide a clear answer to question 2. They show that there are statistically significant racial differences consistent with our earlier visualizations: both AIANa and AIANc are significantly predisposed toward employment in low education fields relative to Whites, with the disparity significantly smaller for the AIANc group.

Do Standard Demographic Factors Account for Occupational Disparity?

Having established that the occupational distribution of AIAN workers differs from that of single-race White workers and is tilted toward low-education fields, we now turn to our third research question: Do standard demographic factors account for the under-representation of AIAN workers in high-education occupations (relative to White workers)? To answer this question, we add additional explanatory variables, beyond race, to the regression framework introduced in the previous section.

Measures of educational achievement are, on the one hand, natural variables to add because of the obvious ties between education attainment and many occupations. On the other hand, using an individual's education to predict whether they are in a high-education occupation may seem circular and thus merits some discussion. To define the dependent variable in our regressions, we classify occupations as high- or low-education based on whether a high or low percentage of incumbents have at least some college education. Thus, on average over the full sample of Whites and AIAN workers, there must be a positive overall average relationship between individual education attainment and whether an individual is in a high- or low-education field. However, it need not automatically be true that each additional level of education will further increase the odds that an individual will hold a high-education occupation. Nor must individual education be related to occupation on average in the AIAN portion of our sample—this population is very small relative to the White portion and thus has little influence on how occupations are ranked. So the race coefficients in a regression of occupational outcome (high- or low-education field) on individuals' race and education can meaningfully show that (holding the effects of individuals' educational attainment constant) AIAN workers are less likely to hold high-education occupations than Whites.

Other factors may also be related to whether a worker has a high-education occupation. For example, compared to jobs in rural areas, proportionately more jobs in metropolitan areas require high levels of education. In Table 6, we show basic summary statistics on the variables we use in our calculations: sex, location in a metropolitan area, presence of an AIAN homeland in the individual's Public Use Microdata Area (HOMELAND in IPUMS; also see Liebler 2010), age, English proficiency, and educational attainment.

Because it is a primary independent variable of interest, we next (Figure 5) show a plot of the educational attainment of AIANa, AIANc, and White workers in 1980, 1990, 2000, and 2010. We see in Figure 5 (next page) that, compared to Whites in each year, a lower proportion of AIAN labor force participants completed each educational level. In 2000 and 2010, AIANa and AIANc are both more highly concentrated in the "high school graduate" category than Whites, with fewer college degrees and greater numbers without high school education. We can also see a general increase in graduation rates for all groups over time. Although AIAN workers are keeping up with overall educational increases, they are not catching up to close the gaps. The AIAN labor force in aggregate is more educated today than in 1980, but they are still less educated than Whites.

FIGURE 5
Educational Attainment of Labor Force Participants by Race, Over Time

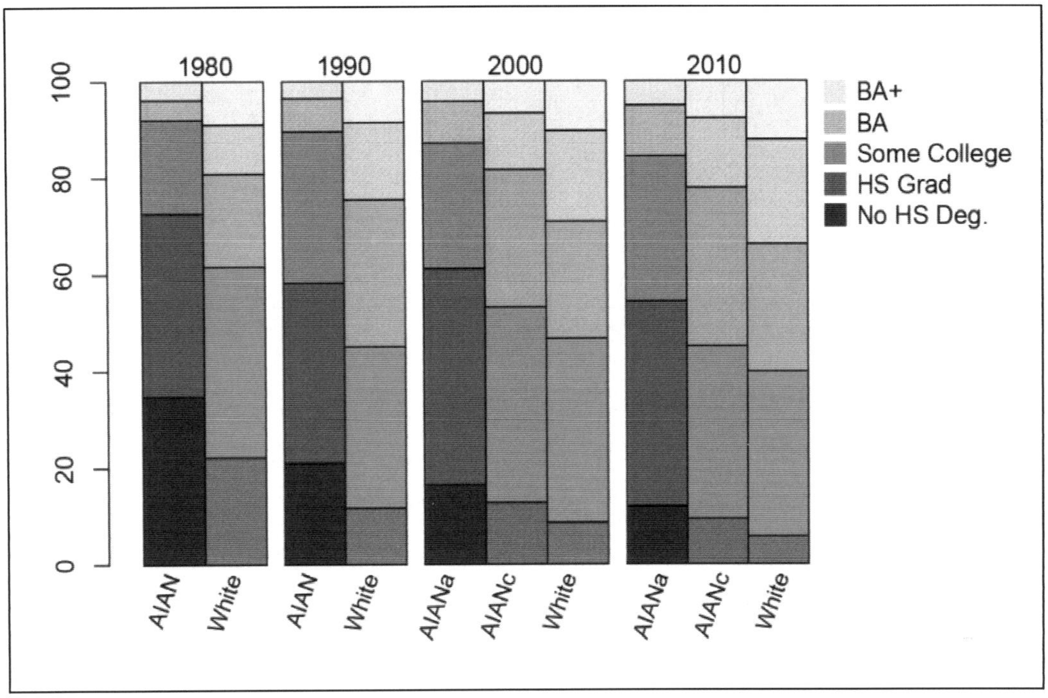

Our predictors of a worker being employed in one of the "highly educated fields" include those shown in Table 6 on the next page, as well as age squared. In Table 7 (two pages hence), we show our results in terms of fitted coefficients, net of interaction effects, with separate columns for men and women as a way of displaying interaction effects.[5] As in Table 5, the coefficients in Table 7 sum to the log-odds of a particular individual being employed in a "highly educated field."

TABLE 6
Summary Statistics for the Year 2010 (Labor Force Participants), Based on Person Weights; the Reported N Is Unweighted

	Race	N	Mean	SD	Min	Max
Female	White	5,397,814	0.4687	0.4990	0	1
	AIANc	45,154	0.4963	0.5000	0	1
	AIANa	54,534	0.4988	0.5000	0	1
Metro Area	White	5,397,814	0.7353	0.4412	0	1
	AIANc	45,154	0.7216	0.4482	0	1
	AIANa	54,534	0.5045	0.5000	0	1
Homeland	White	5,397,814	0.1774	0.3820	0	1
	AIANc	45,154	0.3007	0.4586	0	1
	AIANa	54,534	0.6361	0.4811	0	1
Age	White	5,397,814	42.3989	14.2460	16	95
	AIANc	45,154	39.2964	14.1305	16	94
	AIANa	54,534	39.5640	13.6356	16	94
Not English Proficient	White	5,397,814	0.0049	0.0696	0	1
	AIANc	45,154	0.0041	0.0642	0	1
	AIANa	54,534	0.0060	0.0773	0	1

	Race	N	No HS Deg.	HS Grad.	Some College	BA	BA+
Educational Attainment	White	5,397,814	6%	34%	26%	22%	12%
	AIANc	45,154	9%	36%	33%	14%	8%
	AIANa	54,534	12%	42%	30%	11%	5%

TABLE 7
Fully Adjusted Binomial Regression Predicting Employment in a "Highly Educated Field," for 2010

	Men			Women		
	Estimate	SE		Estimate	SE	
(Intercept)	−4.4954	0.0136	***	−4.3760	0.0147	***
Race "AIANc"	−0.1825	0.0175	***	−0.0944	0.0160	***
Race "AIANa"	−0.2935	0.0186	***	−0.0752	0.0161	***
"Education" HS Grad	1.0581	0.0095	***	1.3205	0.0113	***
"Education" BA	2.8712	0.0097	***	3.0272	0.0115	***
"Education" BA+	4.1782	0.0105	***	4.3453	0.0125	***
Metro Area	0.1517	0.0025	***	0.1517	0.0025	***
Homeland	−0.0171	0.0028	***	−0.0171	0.0028	***
Age	0.0836	0.0005	***	0.0836	0.0005	***
Age^2	−0.0008	0.0000	***	−0.0008	0.0000	***
Not Proficient in English	−0.9483	0.0183	***	−0.9483	0.0183	***

Regardless of race, education is the best predictor of employment in a high education field, and we show in Figure 4 that AIAN workers lag in education relative to White workers. Differences in education between the groups are responsible for a significant amount of the difference between AIAN men's and White men's occupational structure. The same patterns are evident for women. The education coefficients increase sharply with each level of educational attainment for both men and women, and these increases are statistically significant.[6] The effect of education on the odds of working in a highly educated field is stronger for women than for men. Age predicts a maximum probability of high-education employment just above age 50, falling off quadratically. Living in a metropolitan area, not living near a homeland, and being proficient in English also are statistically significant predictors of working in a "highly educated field," although their coefficients show much smaller effects than that of education.

After adjusting for these other factors, including educational attainment, all the race group coefficients are smaller than their values in the previous race-only regression (Table 5). However, they all remain statistically different from zero, implying that the factors we considered did not fully account for the under-representation of AIAN workers in high-education occupations. Compared to the disparities for AIAN men, those between AIAN women and White women are much smaller, or more nearly eliminated, after controlling for our additional factors, but even they remain statistically significant. Thus, the answer to our third research question is "no," at least for the factors we consider.

One difference within the AIAN workforce itself does disappear with our additional controls. The probability of being in a high education occupation is statistically significantly different for AIANa men than for AIANc men, but the difference between AIANa and AIANc women in our full regression model is not significant.[7] We can only say that the fact that the differences for men are substantially smaller in the adjusted model (Table 7) than in the unadjusted model (Table 5) indicates much (though not all) of the observed difference in employment in a high or low education field between AIANa and AIANc male workers is accounted for by additional factors in Table 6.

Conclusion

The raw data on occupational distribution by race reveals a clear disparity between AIAN workers and White workers that has been present since at least 1980. Both AIANa and AIANc workers are underrepresented in high-education fields like management, financial services, and legal professions, relative to White workers. AIAN workers are significantly over-represented in low-education fields like construction, healthcare support, and food preparation. These differences are especially strong when the comparisons are limited to working men.

We find that race-group differences in educational attainment are the single most important explanatory factor behind the race-group differences in whether a worker is in an occupation group with relatively high education in 2010. Accounting for differences in educational outcomes and other factors markedly reduces all the race coefficients relative to their values in a race-only regression, but they are all still statistically different from zero.

These demographic factors also explain much (for men) or all (for women) of the tendency for AIANa workers to be less likely than AIANc workers to work in a highly educated field.

Though AIAN workers have improved their educational attainment in the past decades, White workers have also been increasingly well educated, and a gap remains. Over the same decades, the aggregate occupational dissimilarity of the AIAN workforce seems to have changed little (though data issues prevent us from being certain). Although unmeasured factors also contribute to these occupational dissimilarities, our findings suggest that further efforts to close racial gaps in educational attainment can play an important role in narrowing the occupational dissimilarity between White workers and AIAN workers, thus improving lives and eliminating potential inefficiencies in how jobs are allocated.

Appendix A: Hispanic AIAN

Because the absolute number of Hispanic AIANs is small, we exclude them from the bulk of our analyses. However, in Table 8 on the following page, we show the fitted coefficients for two regression models that include the Hispanic AIAN group (combining Hispanic AIANa with Hispanic AIANc). When compared to White workers, the disparity in education-ranked occupational outcomes is much larger for the Hispanic AIAN group than for either of the non-Hispanic AIAN groups in the "unadjusted" regression, which includes only race/ethnicity explanatory variables. All non-Hispanic AIAN-White disparities are smaller than Hispanic AIAN-White disparities after adjusting for the other covariates, but none are fully accounted for (each coefficient is statistically different from zero). Among AIAN women, the Hispanic AIAN coefficient in the adjusted model is statistically and materially larger than the coefficients for non-Hispanic AIAN workers. Among men, however, the statistical difference between Hispanic AIAN workers and AIANa workers disappears in the adjusted model.

TABLE 8
Unadjusted, and Fully Adjusted, Binomial Regressions Predicting Employment in a "Highly Educated Field," for the year 2010, Including Hispanic AIANs

	Estimate	SE	z value	Pr(>|z|)	
(Intercept)	−0.502	0.001	−564.9	<2e-16	***
Race "AIANc"	−0.400	0.010	−38.86	<2e-16	***
Race "Hispanic AIANa"	−0.948	0.017	−55.23	<2e-16	***
Race "AIANa"	−0.625	0.011	−58.28	<2e-16	***

	Men			Women		
	Estimate	SE		Estimate	SE	
(Intercept)	−4.4942	0.0135	***	−4.3748	0.0147	***
Race "AIANc"	−0.1832	0.0175	***	−0.0953	0.0160	***
Race "Hispanic AIAN"	−0.3189	0.0289	***	−0.2325	0.0274	***
Race "AIAN"	−0.2856	0.0186	***	−0.0681	0.0161	***
"Education" HS Grad	1.0531	0.0095	***	1.3151	0.0113	***
"Education" BA	2.8666	0.0096	***	3.0225	0.0115	***
"Education" BA+	4.1753	0.0105	***	4.3416	0.0124	***
Metro Area	0.1558	0.0025	***	0.1558	0.0025	***
Homeland	−0.0186	0.0028	***	−0.0186	0.0028	***
Age	0.0839	0.0005	***	0.0839	0.0005	***
Age^2	−0.0008	0.0000	***	−0.0008	0.0000	***
Not Proficient in English	−0.9845	0.0179	***	−0.9845	0.0179	***
Mostly Proficient in English	−0.5448	0.0113	***	−0.5448	0.0113	***

Appendix B: Standard Error Estimation

The IPUMS microdata include a *SUBSAMP* variable, indexing all person level observations into 100 representative subsamples of the full data, each 1% of the entire data set. To estimate standard errors for the index of dissimilarity, we calculate the index value on each subsample (X_i) and the entire data set (X), then compute:

$$\mathrm{SE}(X) = \frac{1}{\sqrt{100}}\sqrt{\frac{1}{100}\sum_{i=1}^{100}(X_i - X)^2}$$

Endnotes

[1] We group Hispanic individuals by their ethnicity, regardless of race, and omit them from all race categories, except where explicitly noted. Hereafter, all race groups mentioned are non-Hispanic only unless otherwise indicated.

[2] For the figures, we categorized workers whose industry (not occupation) was reported as "military" to the military category. We then assigned the remaining individuals to their Census-defined occupation group, if any, as recorded by the Census Bureau.

[3] The Karmel and MacLachlan index for individual education groups is also usually equal to or lower than for the general workforce, but here the exception is for the least educated group (those who did not complete high school). The Charles and Grusky index parallels the index of dissimilarity for AIANa workers, except for BA+, where it is not defined due to a zero cell count. For AIANc workers, the Charles and Grusky index for the two extreme education outcomes, No HS and BA+, exceeds the overall index (and is not defined for the BA category).

[4] For the Karmel and MacLachlan index, we find much less dissimilarity for AIANc workers, compared to AIANa workers, in the three lowest education groups but slightly more dissimilarity in the BA and BA+ groups. We cannot produce both AIANa and AIANc values of the Charles and Grusky index for the two college groups, and we find that for the No HS group the AIANc index value exceeds the AIANa index value.

[5] Although it has two columns, this table is for a single model that includes interaction terms that allow race and education effects to be different among men than among women. Other variables are not interacted with sex and thus have the same value in both columns.

[6] We found limited evidence that educational effects differ by race. In a regression (not shown) with race and education interactions, the interactions of race and HS and race and BA were not significant, implying no AIAN-White difference in the effect of education for those levels of attainment. We did find evidence that advanced degrees (BA+) had a somewhat lower effect on occupational outcome for AIANa and AIANc workers than for White workers, but this race effect was small relative to the baseline effect of an advanced degree on workers generally.

[7] Note that AIANc workers are intermediate between Whites and AIANa workers, generally, but this is not true for AIAN women in Table 7. We have no explanation for this difference.

References

Akee, Randall, and Jonathan Taylor. 2014. *Social and Economic Change on American Indian Reservations: A Databook of the US Censuses and the American Community Survey 1990-2010.* http://taylorpolicy.com/us-databook.

Allen, Rebecca, Simon Burgess, Russell Davidson, and Frank Windmeijer. 2015. "More Reliable Inference for the Dissimilarity Index of Segregation." *The Econometrics Journal* 18(1): 40-66.

Alonso-Villar, Olga, Coral Del Rio, and Carlos Gradin. 2012. "The Extent of Occupational Segregation in the United States: Differences by Race, Ethnicity, and Gender." *Industrial Relations* 51(2): 179-212.

Blau, Peter M., and Otis Dudley Duncan. 1967. *The American Occupational Structure.* The Free Press (MacMillan), New York, New York.

Emmel, Alissa, and Theresa Cosca. 2010. "The 2010 SOC: A Classification System Gets an Update." *Occupational Outlook Quarterly* (summer).

Eschbach, Karl. 1992. "Shifting Boundaries: Regional Variation in Patterns of Identification as American Indian." Dissertation (Sociology). Harvard University.

Fox, John. 2003. "Effect Displays in R for Generalised Linear Models." *Journal of Statistical Software* 8(15): 1-27.

Charles, M., and David B. Grusky. 1995. "Models for Describing the Underlying Structure of Sex Segregation." *American Journal of Sociology* 100: 931-971.

Karmel, T., and M. MacLachlan. 1988. "Occupational Sex Segregation—Increasing or Decreasing." *Economic Record* 64: 187-195.

Liebler, Carolyn A. 2010. "Homelands and Indigenous Identities in a Multiracial Era." *Social Science Research* 39: 596-609

Liebler, Carolyn A., Sonya Rastogi, Leticia E. Fernandez, James M. Noon, and Sharon R. Ennis. 2014. "America's Churning Races: Race and Ethnic Response Changes Between Census 2000 and the 2010 Census." United States Census Bureau, Center for Administrative Records Research and Applications, Working Paper 2014-09.

Liebler, Carolyn A., Renuka Bhaskar, and Sonya R. Porter. 2016. "Joining, Leaving, and Staying in the American Indian/Alaska Native Race Category Between 2000 and 2010." *Demography* 53(2): 507-540.

Liebler, Carolyn A., and Andrew Halpern-Manners. 2008. "A Practical Approach to Using Multiple-Race Response Data: A Bridging Method for Public Use Microdata." *Demography* 45(1): 143-55.

Liebler, Carolyn A., and Timothy Ortyl. 2014. "More than One Million New American Indians in 2000: Who are They?" *Demography* 51: 1101-1130.

Lumley, Thomas. 2004. "Analysis of Complex Survey Samples." *Journal of Statistical Software* 9(1): 1-19.

Norris, Tina, Paula L. Vines, and Elizabeth M. Hoeffel. 2012. "The American Indian and Alaska Native Population: 2010." 2010 Census Briefs C2010BR-10.

Office of Management and Budget. 1997. "Revisions to the Standards for the Classification of Federal Data on Race and Ethnicity." Federal Register Notice (October 30).

Passel, Jeffrey S. 1997. "The Growing American Indian Population, 1960–1990: Beyond Demography." *Population Research and Policy Review* 16(1): 11-31.

Ruggles, Steven, Katie Genadek, Ronald Goeken, Josiah Grover, and Mathew Sobek. 2015. Integrated Public Use Microdata Series: Version 6.0 [machine-readable database]. Minneapolis: University of Minnesota.

Taylor, J. 1994. "Measuring the Occupational Segregation of Australia's Indigenous Workforce: A Census-Based Analysis." *Social Indicators Research* 31(2): 175-204.

Watts, Martin. 1998. "Occupational Gender Segregation: Index Measurement and Econometric Modeling." *Demography* 35(4): 489-496.

X. LERA Best Papers I

Does Construction Hazard Awareness Training Reduce Rates of Workers' Compensation Claims?

Ashley L. Schoenfisch
Duke University School of Nursing
Duke University School of Medicine

Clayton Sinyai
CPWR—The Center for Construction Research and Training

Hester Lipscomb
Duke University School of Medicine

Researchers lack consensus on the value of safety training for work-related injury prevention. Within a 9-year (2000–2008) cohort of 17,106 union carpenters in Washington State, this study assessed whether OSHA outreach training was associated with reduced workers' compensation injury claims rates. Training resulted in a 13% reduction (non-significant) in rates of injury. The effect was greater for apprentices and with increasing time since training. Though data do not demonstrate that OSHA outreach training alone decreases injury rates, hazard awareness and protection training should be standard practice in the dangerous construction industry and coupled with more efficient approaches to injury control.

Introduction

Policymakers frequently call for increased investment in job training, but among scholars there is little agreement regarding the efficacy of such training on work-related outcomes. Labor economists seeking to estimate the return on investment from such training find that limitations inherent in the data sources make useful generalizations difficult (Bartel 2000; Dostie 2013). In occupational safety and health, researchers face a similar issue. Practitioners frequently prescribe training as a safety intervention, but a systematic review of the scientific literature by the National Institute for Occupational Safety and Health (NIOSH) concluded that "although the ultimate goal of OHS [occupational health and safety] training is the prevention or reduction of injury, disease and death … the degree of correlation between these outcomes and typical measures of training effects, such as knowledge gain and behavior change, is unclear at best" (Robson et al. 2010).

For workers in the dangerous construction industry, safety training may lead to important increases in worker knowledge and attitudes about occupational safety (Sokas et al. 2009; Williams et al. 2010; Forst et al. 2013) as well as measures of safety behavior (Trabeau et al. 2008; Kaskutas et al. 2013; Ruttenberg 2013). Although evidence supporting safety training's prevention of adverse health outcomes is limited (Cohen et al. 1998; Robson et al. 2010, 2012; Mullan et al. 2015), encouraging examples can be found among plumbers/pipefitters (Kinn et al. 2000), laborers (Dong et al. 2004; Williams et al. 2010), carpenters/drywall tapers (Johnson and Ruppe 2002), residential construction workers (Darragh et al. 2004), and railway construction workers (Bena et al. 2009).

Author's address: Ashley L. Schoenfisch, Duke University School of Nursing, 307 Trent Drive, Durham, North Carolina 27710

When the Occupational Safety and Health Act was established in 1970, the OSHA Training Institute (OTI) established a series of short hazard awareness training courses, including 10-hour (for entry-level workers) and 30-hour (for workers with safety responsibilities) construction hazard awareness courses for trades personnel. The OTI trained federal occupational safety personnel to deliver the courses to federal employees in locations around the nation. The training courses cover the "recognition, avoidance, abatement, and prevention of safety and health hazards in workplaces" and provide "information on workers' rights, employer responsibilities and how to file a complaint" (OSHA 2014).

In the 1990s, OSHA began permitting authorized providers to offer this training in the private sector. Building industry demand drove explosive growth in delivery of "OSHA-10" and "OSHA-30" for construction. Fewer than 20,000 workers per year earned OSHA-10 and OSHA-30 cards in the early 1990s; since 2010, more than 500,000 workers per year have earned one or the other credential. The joint labor–management apprenticeship programs providing vocational training to North America's union-sector construction workers were among the earliest and largest adopters of the training, and today virtually of these programs have incorporated OSHA-10 and/or OSHA-30 into their apprenticeship curricula. Safety-forward gains in construction workers' knowledge and attitudes (Sanyang 2007; Sokas et al. 2009; Wilkins 2011) and behavior (Ruttenberg 2013) have been observed following OSHA-10 training. However, evaluations of the effect of OSHA outreach training on rates of adverse construction work-related outcomes are few (Taylor 2014).

The Pacific Northwest Regional Council of Carpenters (PNWRCC), headquartered in Washington State, reported 20,544 members in 2015. These workers are employed at a defined list of union-signatory contractors; because union-signatory contractors pay into jointly administered benefit trusts at an hourly per capita rate, the union is able to report how many hours each member worked. Some members received OSHA outreach training through the carpenters' affiliated apprenticeship and training programs, while others have not.

Building on an existing, robust, longitudinal cohort of union carpenters in Washington state over the years 2000 through 2008, researchers merged training records to existing, individual-level union records and workers' compensation (WC) claims data to compare the rate and severity of injury claims filed by workers who had received the OSHA outreach training with those who had not.

Methods

As part of prior research, the researchers—with assistance from the Carpenters Trusts of Western Washington (CTWW) and the Washington State Department of Labor and Industries (L&I)—created a cohort of union carpenters examine patterns in work-related injury rates (McCoy et al. 2013). The cohort comprised union eligibility, membership, and WC claims data files, linked at the individual level in a de-identified manner through use of a unique member number. Union eligibility and membership files provided by the CTWW contain for each carpenter: date of birth, gender, monthly hours worked, and data of union entry. The study cohort was limited to individuals who worked at least three months of union hours, with observation beginning in the month eligibility criteria were met (i.e., third month of union work).

WC claims data, provided by L&I contain work-related event details for each claim: date of event, date of claim, claim open/close status, ANSI/OIICS codes for the mechanism of injury, time loss (TL) status, number of paid TL days (with TL compensation paid after the third calendar day of being medically unavailable for work, not including the day of injury), and associated costs. A TL claim is one in which there was not approved work available for the injured worker within the medical restrictions; this categorization was provided to the researchers by L&I. Cost estimates are limited to payment types reported consistently between state fund and self-insured claims (i.e., accident fund dollars). This approach improves the consistency of costs across payers, but it excludes the medical portion of claims costs. All costs are adjusted using the consumer price index to 2015 dollars and assume all dollars were paid on the date of injury. Data from the claim pool, extracted in April 2010 (McCoy et al. 2013), were updated April 2016 to reflect L&I's current methodology and estimates related to paid TL and costs.

In the current study, these files were linked at the individual level to a union carpenter training data file in a similar manner. The training file, provided by the CTWW, is composed of individual-level data on the dates of OSHA-10, OSHA-30, and OSHA-500 level courses completed through PNWRCC-affiliated training centers. For purposes of this study, an individual carpenters' union work hours were categorized as "trained" if the carpenter completed an OSHA outreach training course within the prior five years.

The characteristics of union carpenters and worker-hours with and without OSHA outreach training, as well as by type of OSHA outreach training course, were described. Time at risk (based on hours of work) and work-related outcomes were stratified by categories of OSHA outreach training (yes/no), age, gender, time in the union, and calendar time. Age and time in the union were treated as time-varying variables. Injury rates (# events/200,000 hours of union work), unadjusted and adjusted rate ratios, and 95% confidence intervals (CI) were calculated using Poisson regression, with the natural log of hours worked as the offset variable (Nizam 2014) and scaling to correct for overdispersion appropriately (Pearson $\chi^2/$ degrees of freedom > 1.5). For analyses of TL days and costs (outcomes that are highly skewed due to a significant proportion of the cohort with no TL days or costs), negative binomial models (Hilbe 2011) were used to examine outcome rates, rate ratios, and 95% CIs, with generalized estimating equations (GEE) (Zeger et al. 1988) to account for within-subject correlation.

For this training intervention, it is plausible to assume an enhanced effect immediately following the training class and diminishing with time. It is also plausible to observe a gradually increased effect over time as trained workers spend more time in the field gaining experience applying training-based knowledge. To explore various scenarios, successive "lagged" models were constructed to examine rate ratios using varying windows of time following training (Figure 1).

FIGURE 1
Lagging Approach

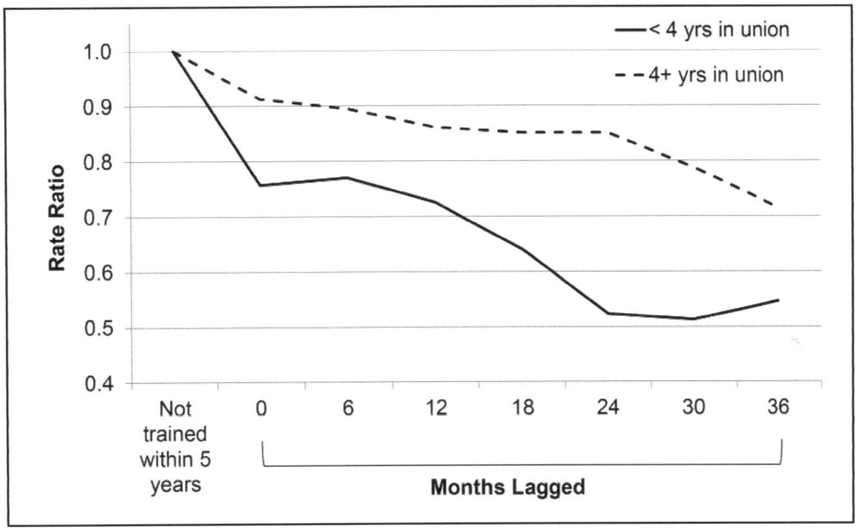

Study procedures were approved by the Institutional Review Boards of Duke University School of Nursing and the Washington State Department of Health and Human Services.

Results

The study cohort is composed of 17,106 carpenters who worked 99,411,000 union hours in Washington state between 2000 and 2008. Nearly all are male (n = 16,466; 97.7%). The average carpenter age increased from 41.3 years in 2000 to 44.9 in 2005, and down to 42.0 in 2008 (McCoy et al. 2013). A total of 9,886 WC claims were reported and accepted for the cohort, 18.6% (n = 1,841) of which were TL claims. The rate of injuries overall and TL injuries declined 42% and 47%, respectively over the study period. A total of 2,910 (29.2%) injuries were characterized as a "Fatal Four" mechanism: fall from elevation (5.0%), electrocutions (0.3%), struck by (21.3%), and caught in/between (2.8%).

During the nine-year study period, 1,017 union members took 1,089 OSHA outreach training courses through the PNWRCC-affiliated training centers. Of these, 7% (n = 67) took more than one OSHA outreach training course. Most of the courses taken were "OSHA-10 Construction" (85%), followed by "OSHA-30 Construction" (10%) and "OSHA-500 (or OSHA-502) Construction" (3%). The number of completed courses was generally steady over time, averaging 135 union members completing courses per year, with the exception of a noticeable drop in 2005/2006 to an average of 73 members completing courses per year, a 46% decrease. (In comparison, the average number of union hours worked per year during 2005 and 2006 were 8% lower than that observed in other years.)

When the individual-level training data were merged to the study cohort of members who contributed at least three months of union hours, 57.5% (n = 585) of these union carpenters who took OSHA outreach training during 2000–2008 *did not link* to the study cohort, likely due to the cohort being limited to carpenters with at least three months of union hours. Analyses related to the characteristics of workers taking OSHA outreach training, as well as those related to the effectiveness of OSHA outreach training, are based on training data from the 432 carpenters who linked to the study cohort. They completed 464 OSHA outreach training classes. Most OSHA outreach training took place in carpenters' early years in the union, and particularly in the first year for carpenters completing OSHA-10 training (Table 1). Over 78% of OSHA-10 training was among carpenters who were likely in apprenticeship programs (i.e., more than four years in the union).

TABLE 1
Stratified by Years in Union, OSHA Outreach Training Courses Completed by Union Carpenters Through PNWRCC Training Centers, 2000–2008

	OSHA-10 (n = 377)		**OSHA-30 (n = 76)**	
Years in union	n	%	n	%
<1	250	(66.3)	26	(34.2)
1 to <2	31	(8.2)	4	(5.3)
2 to <4	15	(4.0)	6	(7.9)
4 to <6	16	(4.2)	7	(9.2)
6 to <8	9	(2.4)	7	(9.2)
8 to <10	14	(3.7)	3	(3.9)
10 to <20	24	(6.4)	12	(15.8)
20+	18	(4.8)	11	(14.5)

Rates of reported and accepted WC claims for work-related injury were 17.4 per 200,000 union hours worked among "trained" members, compared to 19.9 per 200,000 union hours worked among "non-trained" members ($RR_{\text{Trained vs. Non-Trained}}$ 0.87, 95% CI 0.72–1.06) (Table 2). This relative difference did not vary significantly across categories of age or calendar time. The protective effect was more apparent among carpenters in their apprenticeship years (more than four years in the union; RR 0.76, 95% CI 0.54–1.07). No significant difference was observed in the proportion of injuries resulting in paid TL days (trained: 21.2%; non-trained: 18.6%), the median number of paid TL days per TL injury (trained: 105; non-trained: 92), or the rate of TL injury by training status (RR 1.00, 95% CI 0.65–1.53). Unadjusted and adjusted analyses yielded similar findings; the unadjusted estimates are presented.

TABLE 2
Injury Frequencies, Time at Risk, Rates, Rate Ratios, and 95% CI of WC Claims for Injury by OSHA Outreach Training, Overall and by Worker Age, Time in Union, Type of Work, and Calendar Year

	OSHA outreach training in past 5 years									
	Yes				No					
	Hours	# Injuries	Rate[a] (95% CI)		Hours	# Injuries	Rate[a] (95% CI)		Rate ratio[b] (95% CI)	
Injuries, Overall	1,138,147	99	17.4	(14.3–21.2)	98,272,853	9,787	19.9	(19.5–20.3)	0.87	(0.72–1.06)
By carpenter age										
<30 years	269,493	26	19.3	(13.1–28.3)	16,423,237	1,820	22.2	(21.2–23.2)	0.87	(0.59–1.28)
30 to <40 years	356,751	27	15.1	(10.4–22.1)	28,177,004	2,960	21.0	(20.3–21.8)	0.72	(0.49–1.05)
40+ years	506,276	45	17.8	(13.3–23.8)	53,538,101	4,999	18.7	(18.2–19.2)	0.95	(0.71–1.28)
By time in the union										
Apprentices (<4 years)	355,427	33	18.6	(13.2–26.1)	23,586,374	2,890	24.5	(23.6–25.4)	0.76	(0.54–1.07)
Journeymen (≥4 years)	782,720	66	16.9	(13.3–21.5)	74,686,479	6,897	18.5	(18.0–18.9)	0.91	(0.72–1.16)
By calendar year										
2000–2002	74,999	13	34.7	(20.1–59.7)	31,806,516	3,806	23.9	(23.2–24.7)	1.45	(0.84–2.50)
2003–2005	347,504	29	16.7	(11.6–24.0)	27,597,535	2,713	19.7	(19.0–20.4)	0.85	(0.59–1.22)
2006–2008	715,644	57	15.9	(12.3–20.7)	38,868,803	3,268	16.8	(16.3–17.4)	0.95	(0.73–1.23)

[a] Events per 200,000 work hours

[b] Compares rate of injury among trained versus rate of injury among non-trained

The number of paid TL days per 200,000 hours worked was lower among trained carpenters (trained: 547 per 200,000 union hours worked; non-trained: 1,752 per 200,000 union hours worked). Similar patterns were observed for rates of associated costs (trained: $0.79 per union work hour; non-trained: $1.92 per union work hour). Efforts were made to examine rates and relative rate differences of paid lost days and costs between trained versus non-trained workers using negative binomial regression and GEE. However, models often did not converge, and resulting estimates were unstable.

In lagged analyses, an increasingly protective effect of training was observed with increasing time since training, although results are statistically non-significant (e.g., $RR_{36\text{-month lag}}$ 0.66, 95% CI 0.35–1.25). This protective effect was greater among carpenters with more than years in the union, compared to their more experienced counterparts (Figure 2).

FIGURE 2
Rate Ratios Comparing Work-Related Injury Rates Surrounding OSHA Outreach Training, Lagged by Increasing Amounts of Time Post-Training and Stratified by Years in the Union

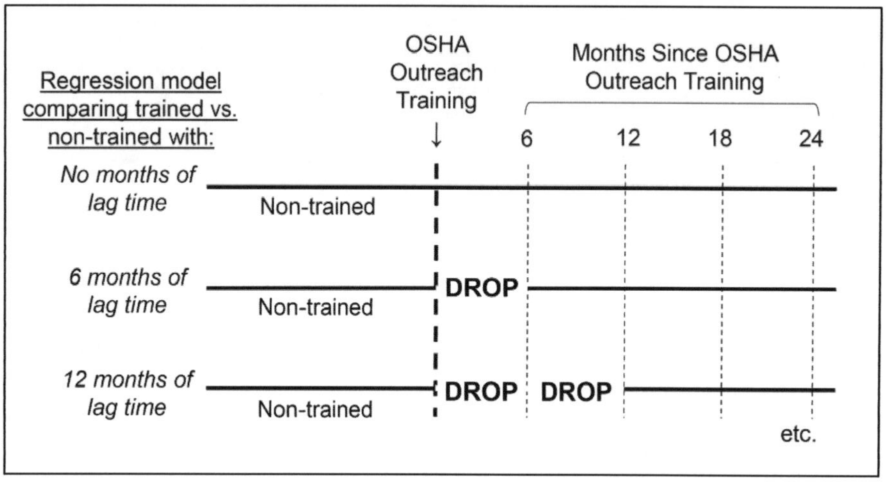

Discussion

Effect of OSHA Outreach Training on Rates of Injury and Related Outcomes

Participation in OSHA outreach training was associated with a 13% reduction in rates of injury overall (measured by WC claims filed and accepted), with a more pronounced effect among workers during their apprenticeship years. Although these findings were not statistically significant and were attenuated in adjusted analyses, they bore strong similarities to patterns observed by Dong et al. (2004) who evaluated the effect of safety training on rates of work-related injury among union laborers in Washington state over two years in the mid-1990s. The authors saw a 12% reduction in injury rates among trained versus untrained workers, and this effect was greater among younger workers—a finding the authors attributed to less informal on-the-job training and a greater propensity of younger and less experienced workers to benefit from injury prevention training material (Dong et al. 2004). Findings from both studies suggest that workers with the least experience receive the most benefit from safety training. Appropriately, the Carpenters' construction apprenticeship program mandates OSHA-10 outreach training for incoming apprentices, and first-year union members accounted for a majority of the OSHA-10 students during the period under study.

Lagged analyses suggested that as time passed since training, the magnitude of the protective effect of training increased. It is not clear whether this is attributable to the improved application of classroom knowledge with increased on-the-job experience; a sorting process in the labor market in which the safest firms seek out safety-trained workers, and vice versa; or some combination of these and other unidentified factors.

No difference was observed in the proportion of injuries resulting in paid TL, median number of paid TL days per TL injury, or rates of TL injury by training status. However, the rates of paid TL days and costs were lower among trained workers. The data were not robust enough to allow exploration of this finding beyond a comparison of simple rates, so the results are open to multiple interpretations. OSHA outreach training may indeed effectively reduce rates of severe work-related injury events; it may also be the case that firms with a preference for employees with this safety credential are also more likely to employ controls to abate severe jobsite hazards. (The nature of the WC claims data available precluded identification or follow-up of firms with the sharply reduced WC claims costs.)

An important body of occupational safety and health research indicates another possibility: workers with OSHA outreach training may exhibit different injury reporting behavior than those without. Measuring WC claims activity is not the same as measuring occupational injuries. Past research has documented that a large number of workplace injuries are not reported to WC often by employees who find the process burdensome or fear employer retaliation (Azaroff et al. 2002; Shannon and Lowe 2002; Leigh et al. 2004; Fan et al. 2006; Rosenman et al. 2006; Friedman and Forst 2007; Welch et al. 2007; Lipscomb et al. 2013, 2015; Wuellner and Bonauto 2014). In one study, 30% of carpenter apprentices completing confidential surveys said that adverse workplace events were "rarely" or "almost never" reported (Lipscomb et al. 2013). In another, more than one-quarter of plumbing and pipefitting journeymen acknowledged suffering a workplace injury they did not report (Taylor Moore et al. 2013). Firms with high-quality safety and health management programs stress reporting of all workplace injuries, even minor ones; OSHA outreach training includes instruction on reporting injuries and on workers' legal protections against retaliation under OSHA. These incentives would tend to increase the propensity of workers to report minor injuries, in which case the 13% reduction in WC claims associated with OSHA outreach training observed in this study would conceal an even larger gap in real-world injury outcomes.

Implications for the Construction Labor Market

Epidemiologists and other occupational safety and health specialists typically use WC claims data to define occupational injury events and conduct related analyses. Managers and owners, on the other hand, have an intrinsic interest in WC claims data—and particularly costs associated with claims—as a reflection of liability. Even assuming a large number of unreported injuries, injury liability remains a major cost center for the construction industry, with base WC premiums in 2016 costing an average base rate of $2.7466 per hour in Washington state, or $5,493 per worker per year (Washington State Department of Labor and Industries 2016). It is plausible that funds spent to reduce those costs will bring a better return on investment than an equal amount invested to improve labor productivity.

In this study, if taken at face value, the training delivered an astronomical return on investment. While OSHA outreach training for these carpenters is wrapped into their general apprenticeship program, and not priced separately, the Associated General Contractors of Washington advertised a public rate of $175 per worker for OSHA-10 for construction during 2016 (AGC Education Foundation 2016)—an annualized cost of $35 if the benefits are held to the past five years. With untrained workers generating $0.79 per work hour in injury claims, and untrained workers generating $1.92, training would appear to save the firm $1.13/hour. That would indicate $2,260 in annual savings for a full-time employee (2,000 hours)—for a net gain of $2,225 on a $35 training investment! Is this the hiring hall? Send me more of those carpenters with an OSHA-10 or OSHA-30 card in their pocket!

And in some ways, using individual WC claims as a proxy for workplace injury understates overall compensation claims savings. The benefits of effective safety training have the potential to reach more than the worker involved. The trained worker may be less likely to cause injuries to others, or they may be more likely to speak up when they see unsafe practices or working conditions (Dong et al. 2004; Williams et al. 2010; Forst et al. 2013; Ruttenberg 2013). For example, in Forst et al. (2013) trained workers described making other workers aware of hazards, talking with supervisors about hazards, and taking on informal safety leadership roles. This potential "public good" created by training, though not captured using data from this study, could reduce the magnitude of a protective effect of OSHA outreach training on trainees' WC claims rates.

The researchers do not contend that firms should expect to realize such a large return on their safety training investment. Other factors were likely at work here. Further, occupational safety and health practitioners recommend that firms address hazards using a hierarchy of controls (Castillo et al. 2011), attempting to reduce exposure to hazards—and ultimately work-related injuries—by eliminating a hazard or substituting safer materials and practices.

Leaving the hazard in place and training workers to avoid it is not an effective or recommended solution. Firms with a preference for hiring workers with OSHA outreach training have likely invested more than their peers in these preferred, more effective controls—and it may be that workers with OSHA outreach training also recognize and prefer employers who approach hazard control in this manner.

Limitations and Strengths

Our statistical analysis was constrained by the number of trained workers we were able to identify. Presumably, a substantial number of journeyman carpenters in this cohort received OSHA outreach training prior to the period under study, but records for that time were unavailable. Additionally, information on carpenters' OSHA outreach training through other means (e.g., other training centers, online courses) was not available. Although the training records indicated that 1,017 workers received OSHA outreach training during the period, fewer than half of these workers met the cohort requirement of at least three months of union work. As most of the training was delivered to workers with fewer than four years in the union, this is likely a consequence of apprentice attrition (Bilginsoy 2003, Byrd and Weinstein 2005, Glover and Bilginsoy 2005). Our analysis was therefore limited to comparing 432 workers who received the training during the study period to the others.

This study demonstrates both the utility and importance of robust, longitudinal surveillance data in the appraisal of long-term evaluations of the effectiveness of workplace safety interventions, as well as challenges in clearly evaluating training effects in this dangerous industry. Similar exploration of construction workers' training and injury experience outside of this dynamic cohort is unrealistic, if not impossible, in the open shop environment or otherwise. Time at risk and reported injuries were well defined, allowing for the calculation of injury rates and their examination over time surrounding documented training.

Conclusion

These data do not demonstrate the likelihood that OSHA outreach training alone meaningfully decreases injury rates among union construction industry workers. Worker training in hazard awareness, injury prevention and mitigation, workers' rights, and how to report adverse work-related events *should* be standard practice, but such training alone cannot be expected to prevent work-related injury to a significant degree in this dangerous industry. Training focusing on informing works in hazard awareness and protection does not address organizational issues that broadly influence worker safety (Dekker et al. 2013). For public health researchers, the lesson here is a familiar one: Training is a supplement, not a substitute, for more effective (but sometimes more costly) strategies on occupational health's "hierarchy of controls"—eliminating a hazard, substituting a safer material or technique, or abating a hazard through engineering solutions. Firms and economists, however, may find other lessons.

Acknowledgments

We thank Thomas Barrett and Jeff Jewell at the PNWRCC Training Centers and Lin Conley and Randy Parker at the CTWW for their support and access to data on carpenters' completion of OSHA outreach training courses. We also thank David Bonauto and Darrin Adams at the Safety and Health Assessment and Research for Prevention Program in the Washington State Department of L&I for support of the project, provision of L&I data, and thoughtful comments on results. This report was made possible by CPWR—The Center for Construction Research and Training, through cooperative agreement number U60-OH009762 from NIOSH. Its contents are solely the responsibility of the authors and do not necessarily represent the official views of the CPWR or NIOSH.

References

AGC Education Foundation. 2016. "Class Highlights." constructionfoundation.org

Azaroff, L. S., et al. 2002. "Occupational Injury and Illness Surveillance: Conceptual Filters Explain Underreporting." *American Journal of Public Health*, Vol. 92, no. 9, pp. 1421-1429.

Bartel, A. 2000. "Measuring the Employer's Return on Investments in Training: Evidence from the Literature." *Industrial Relations*, Vol. 39, no. 3, pp. 502-524.

Bena, A., et al. 2009. "Effectiveness of the Training Program for Workers at Construction Sites of the High-Speed Railway Line between Torino and Novara: Impact on Injury Rates." *American Journal of Industrial Medicine*, Vol. 52, no. 12, pp. 965-972.

Bilginsoy, C. 2003. "The Hazards of Training: Attrition and Retention in Construction Industry Apprenticeship Programs." *Industrial & Labor Relations Review*, Vol. 57, no. 1, pp. 54-67.

Byrd, B., and M. Weinstein. 2005. "Construction Apprenticeship in Oregon: An Analysis of Data on Union and Open-Shop Apprenticeship Programs." Report prepared for the Oregon State Building and Construction Trades Council, Portland, OR.

Castillo, D., et al. 2011. "Injuries and Occupational Safety." In Barry S. Levy, David H. Wegman, Sherry L. Baron, and Rosemary K. Sokas, eds., *Occupational and Environmental Health: Recognizing and Preventing Disease and Injury*. Philadelphia, PA: Lippincott, Williams, and Wilkins.

Cohen, A., et al. 1998. "Assessing Occupational Safety and Health Training." Cincinnati, PH: National Institute for Occupational Safety and Health: 98-145.

Darragh, A. R., et al. 2004. "Effectiveness of the Homesafe Pilot Program in Reducing Injury Rates Among Residential Construction Workers, 1994–1998." *American Journal of Industrial Medicine*, Vol. 45, no. 2, pp. 210-217.

Dekker, S. W., et al. 2013. "The Little Engine Who Could Not: 'Rehabilitating' the Individual in Safety Research." *Cognition, Technology & Work*, Vol. 15, no. 3, pp. 277-282.

Dong, X., et al. 2004. "Effects of Safety and Health Training on Work-Related Injury among Construction Laborers." *Journal of Occupational and Environmental Medicine*, Vol. 46, no. 12, pp. 1222-1228.

Dostie, B. 2013. "Estimating the Returns to Firm-Sponsored On-the-Job and Classroom Training." *Journal of Human Capital*, Vol. 7, no. 2, pp. 161-189.

Fan, Z. J., et al. 2006. "Underreporting of Work-Related Injury or Illness to Workers' Compensation: Individual and Industry Factors." *Journal of Occupational and Environmental Medicine*, Vol. 48, no. 9, pp. 914-922.

Forst, L., et al. 2013. "More Than Training: Community-Based Participatory Research to Reduce Injuries Among Hispanic Construction Workers." *American Journal of Industrial Medicine*, Vol. 56, no. 8, pp. 827-837.

Friedman, L. S., and L. Forst. 2007. "The Impact of OSHA Recordkeeping Regulation Changes on Occupational Injury and Illness Trends in the US: A Time-Series Analysis." *Occupational and Environmental Medicine*, Vol. 64, no. 7, pp. 454-460.

Glover, R. W., and C. Bilginsoy. 2005. "Registered Apprenticeship Training in the US Construction Industry." *Education+ Training*, Vol. 47, no. 4/5, pp. 337-349.

Hilbe, J. M. 2011. *Negative Binomial Regression*. New York, NY: Cambridge University Press.

Johnson, K. A., and J. Ruppe. 2002. "A Job Safety Program for Construction Workers Designed to Reduce the Potential for Occupational Injury Using Tool Box Training Sessions and Computer-Assisted Biofeedback Stress Management Techniques." *International Journal of Occupational Safety and Ergonomics*, Vol. 8, no. 3, pp. 321-329.

Kaskutas, V., et al. 2013. "Fall Prevention and Safety Communication Training for Foremen: Report of a Pilot Project Designed to Improve Residential Construction Safety." *Journal of Safety Research*, Vol. 44, no. February, pp. 111-118.

Kinn, S., et al. 2000. "Evaluation of Safety Orientation and Training Programs for Reducing Injuries in the Plumbing and Pipefitting Industry." *Journal of Occupational and Environmental Medicine*, Vol. 42, no. 12, pp. 1142-1147.

Leigh, J. P., et al. 2004. "An Estimate of the US Government's Undercount of Nonfatal Occupational Injuries." *Journal of Occupational and Environmental Medicine*, Vol. 46, no. 1, pp. 10-18.

Lipscomb, H., et al. 2013. "Safety, Incentives, and the Reporting of Work-Related Injuries Among Union Carpenters: 'You're Pretty Much Screwed If You Get Hurt at Work.'" *American Journal of Industrial Medicine*, Vol. 56, no. 4, pp. 389-399.

Lipscomb, H., et al. 2015. "Non-Reporting of Work Injuries and Aspects of Jobsite Safety Climate and Behavioral-Based Safety Elements among Carpenters in Washington State." *American Journal of Industrial Medicine*, Vol. 58, no. 4, pp. 411-421.

McCoy, A., et al. 2013. "20 Years of Work-Related Injury and Illness Among Union Carpenters in Washington State." *American Journal of Industrial Medicine*, Vol. 56, no. 4, pp. 381-388.

Mullan, B., et al. 2015. "Active Behaviour Change Safety Interventions in the Construction Industry: A Systematic Review." *Safety Science*, Vol. 79, no. November, pp. 139-148.

Nizam, A. 2014. "Poisson Regression Analysis." In David G. Kleinbaum, Lawrence L. Kupper, Azhar Nizam, and Eli S. Rosenberg, eds., *Applied Regression Analysis and Other Multivariable Methods*. Boston, MA: Cengage Learning.

OSHA. 2014. "OSHA Outreach Training Program." www.osha.gov/dte/outreach

Robson, L. S., et al. 2010. "A Systematic Review of the Effectiveness of Training & Education for the Protection of Workers." Toronto: Institute for Work & Health, 2010; Cincinnati, Oh: National Institute for Occupational Safety and Health. (This publication can also be tracked as DHHS (NIOSH) Publication No. 2010-127.)

Robson, L. S., et al. 2012. "A Systematic Review of the Effectiveness of Occupational Health and Safety Training." *Scandinavian Journal of Work, Environment & Health*, Vol. 38, no. 3, pp. 193-208.

Rosenman, K. D., et al. 2006. "How Much Work-Related Illness and Injury Is Missed by the Current National Surveillance System?" *Journal of Occupational and Environmental Medicine*, Vol. 48, no. 4, pp. 357-365.

Ruttenberg, R. 2013. "The Economic and Social Benefits of OSHA-10 Training in the Building and Construction Trades." CPWR. www.cpwr.com/publications/economic-and-social-benefits-osha-10-training-building-and-construction-trades

Sanyang, A. 2007. "OSHA-10: Key in Construction Safety Awareness." *Blueprints: A Technical Publication of American Society of Safety Engineers' Construction Practice Specialty*, Vol. 7, no. 1, pp. 12-17.

Shannon, H. S., and G. S. Lowe. 2002. "How Many Injured Workers Do Not File Claims for Workers' Compensation Benefits?" *American Journal of Industrial Medicine*, Vol. 42, no. 6, pp. 467-473.

Sokas, R. K., et al. 2009. "An Intervention Effectiveness Study of Hazard Awareness Training in the Construction Building Trades." *Public Health Reports*, Vol. 124, no. 4, pp. 161-168.

Taylor, E. L. 2014. "Safety Benefits of Mandatory OSHA-10 Hour Training." CPWR. www.cpwr.com/publications/safety-benefits-mandatory-osha-10-hour-training

Taylor Moore, J., et al. 2013. "Construction Workers' Reasons for Not Reporting Work-Related Injuries: An Exploratory Study." *International Journal of Occupational Safety and Ergonomics*, Vol. 19, no. 1, pp. 97-105.

Trabeau, M., et al. 2008. "A Comparison of 'Train-the-Trainer' and Expert Training Modalities for Hearing Protection Use in Construction." American Journal of Industrial Medicine, Vol. 51, no. 2, pp. 130-137.

Washington State Department of Labor and Industries. 2016. "Rates for Workers' Compensation: 2016 Base Rates by Business Type and Classification Code." www.lni.wa.gov/Claimsins/Insurance/Ratesrisk/Check/Rateshistory/Default.Asp

Welch, L., et al. 2007. "Is the Apparent Decrease in Injury and Illness Rates in Construction the Result of Changes in Reporting?" *International Journal of Occupational and Environmental Health*, Vol. 13, no. 1, pp. 39-45.

Wilkins, J. R. 2011. "Construction Workers' Perceptions of Health and Safety Training Programmes." *Construction Management and Economics*, Vol. 29, no. 10, pp. 1017-1026.

Williams, Q., et al. 2010. "The Impact of a Peer-Led Participatory Health and Safety Training Program for Latino Day Laborers in Construction." *Journal of Safety Research*, Vol. 41, no. 3, pp. 253-261.

Wuellner, S. E., and D. K. Bonauto. 2014. "Exploring the Relationship Between Employer Recordkeeping and Underreporting in the BLS Survey of Occupational Injuries and Illnesses." *American Journal of Industrial Medicine*, Vol. 57, no. 10, pp. 1133-1143.

Zeger, S., et al. 1988. "Models for Longitudinal Data: A Generalized Estimating Equation Approach." *Biometrics*, Vol. 44, no. December, pp. 1049-1060.

"Sure the Snack Bar Is Great, But a Union Contract Would Be Even Sweeter!": Union Organizing in the Digital-Only News Media

Howard R. Stanger
Wehle School of Business
Canisius College

> As digital journalism finds its place in the new-media landscape, helped by a crop of web-only publications, younger journalists are beginning to demand the kind of work protections, decent wages, and newsroom solidarity that many of their older counterparts once enjoyed
> —Maharidge, 3/2/16

In January 2015, *Washington Post* reporter Lydia DePillis hypothesized about why unions have failed to organize workers in new media.[1] First, she contends, they have lost leverage because the labor supply of aspiring journalists exceeds labor demand in an environment where content is becoming increasingly commoditized. Second, the mostly young (white), educated workers from comfortable backgrounds are less familiar with unions who've "built personal brands" that they can transfer to other media companies (DePillis, 1/30/15).

Although partially correct, she was proven wrong by the events that transpired six months later when digital journalists at *Gawker* chose the Writers Guild of America, East (WGAE) to represent them. Eighty out of 107 journalists signed authorization cards in support of WGAE. Management voluntarily recognized the union without a fight. In the four months following this surprising victory, journalists at six other digital-only newsrooms chose either the WGAE or the NewsGuild-CWA (NG), formerly The Newspaper Guild-CWA to represent them.[2] All except one property voluntarily recognized the unions, another unexpected outcome[3] (Table 1, next page).

This paper explores this recent wave of union organizing in digital newsrooms in the United States and attempts to answer the following questions: What issues have spurred labor organizing? Why was the road to unionization paved smoothly? What is the current state of collective bargaining? What are the future prospects for unions in online news and for labor in general? Finally, what, if any, policy implications emerge from this study?

Troubles in the Fourth Estate: Legacy Newspapers and the Shift to Digital Media

Since 2005, the print newspaper industry has experienced a dramatic decline in print advertising revenues, its lifeblood. Newspapers posted a record high $49.4 billion in advertising revenues that year, but dropped precipitously until reaching $19.9 billion in 2014, the last year for which industrywide data is available.[4] Estimates for 2015 indicate an addition drop of 8%, the biggest single-year drop since 2009.[5] Most of the decline continues to be in print ads, which still comprise 75% of total revenues. Digital advertising revenues also fell 2% in 2015. By comparison, total Internet revenues grew over 20%, to $59.6 billion, last year (Mitchell and Page, March 2014, p. 7; Mitchell et al., June 2016, p. 9, 13; Sass, 6/27/16).

Author's address: Department of Management, Wehle School of Business, Canisius College, Buffalo, NY 14208

TABLE 1
Digital-Only News Outlets and Union Situation

News Outlet	Year Founded	Principal Owner	Location	Union	Date of Recognition	Rep Process	Vote or CBU Size	Status
Gawker	2003	Nick Denton	NYC	WGAE	6.3.2015	VR*	80-27	CBA (3.1.16)
Guardian US	2011	Guardian UK	NYC	NewsGuild	7.29.2015	VR	45	Negot.
Salon	1995	Salon Media	SF & NYC	WGAE	8.1.2015	VR	26-0	Negot.
VICE	1994	Shane Smith	NYC	WGAE	8.10.2015	VR	75	CBA (4.21.16)
ThinkProgress	2005	Ctr for Am Progress	Wash DC	WGAE	9.30.2015	VR	30	Negot.
Al Jazeera America	2013	Qatari Govt	NYC	NewsGuild	10.5.2015	NLRB	32-5	Closed
Huffington Post	2005	Verizon/AOL	NYC	WGAE	1.14.2016	VR	262	Negot.

*Voluntary recognition

Troubles in the Fourth Estate: Legacy Newspapers and the Shift to Digital Media

Since 2005, the print newspaper industry has experienced a dramatic decline in print advertising revenues, its lifeblood. Newspapers posted a record high $49.4 billion in advertising revenues that year, but dropped precipitously until reaching $19.9 billion in 2014, the last year for which industrywide data is available.[4] Estimates for 2015 indicate an addition drop of 8%, the biggest single-year drop since 2009.[5] Most of the decline continues to be in print ads, which still comprise 75% of total revenues. Digital advertising revenues also fell 2% in 2015. By comparison, total Internet revenues grew over 20%, to $59.6 billion, last year (Mitchell and Page, March 2014, p. 7; Mitchell et al., June 2016, p. 9, 13; Sass, 6/27/16).

The largest losses have occurred in the print classified area, where ad dollars have migrated to digital and captured by companies such as Craigslist, Facebook, Google, and other pure-play Internet companies. In 2014, for example, digital media captured 77% of the $31.6 billion classified revenues in the United States. The effect on newspapers has been financially devastating: classified ad revenues dropped from almost $20 billion in 2000 to $4.1 billion in 2013.

Despite a move to digital media, where a record-setting 166 million people consumed online newspaper content in October 2014, digital advertising still represents a small share (14.5%) of total newspaper ad revenues. Newspapers' share of the digital advertising market fell from 16.4% in 2003 to 7.9% in 2013, while total advertising and audience revenues across all categories shrank by over 35% between 2003 and 2013, from $57.4 billion to $37.6 billion. A major contributing factor is that the industry has lost nearly half its print readership since 2005[6] (Mutter, 7/16/14 & 11/20/14; Mitchell and Page, March 2014, p. 7 and April 2015, pp. 2-4; Sass, 11/25/14).

Although still profitable,[7] pre-tax profits for publicly traded newspaper companies tanked by 37% between 2003 and 2013. Over the past two decades, major U.S. newspapers have seen multiple ownership changes and steep declines in value. This has occurred within the past few years in Boston, Philadelphia, Chicago, Washington, San Diego, Minneapolis, and elsewhere. Industry changes also have included spinoffs of newspaper units into separate companies and consolidation. Between 2004 and 2014, the number of dailies fell from 1,457 to 1,331 (Mutter, 7/16/14; Mitchell, 5/22/15; Mitchell et al., June 2016, pp. 18-19).

The economic difficulties in print newspapers have led to major contractions in newsroom staffing. Since a recent high of 56,400 jobs in 2001, employment fell to 32,900 by 2015, a 41% decline. While major staffing cuts occurred between 2015 and 2016, most occurred between 2005 and 2010 and were felt largely by the mid-size urban metros. Employment, however, grew at smaller papers, while overall employment levels have stabilized (American Society of News Editors, 7/28/15; Fox, 7/28/15; Mitchell, June 2016, p. 17). These employment trends have landed the job of newspaper reporter high on *CareerCast's* annual list of endangered jobs. Since 2013, it has vied with lumberjack as the worst job (Mullin and Hare, 10/14/15).

According to the U.S. Bureau of Labor Statistics, total newspaper employment dropped by about 60% since 1990, going from 458,000 to 183,000. By comparison, employment in Internet publishing and online broadcasting rose from about 30,000 to almost 198,000, with former taking off since 2008 (Greenslade, 6/6/2016). The economic challenges facing the newspaper industry have had a direct impact not only on employment but also labor relations.

Stanger (2013) provides a detailed overview of these linkages. He found that the biggest developments in labor relations in the twenty-first century have been the decline of union membership, concessionary and hard bargaining, and unions fighting to protect the integrity of their contracts. While difficult and challenging business conditions and labor relations affect traditional print newspapers, there has been recent and rapid growth in the digital-only media segment.

Digital News: A Rising Phoenix or Flash in the Pan?

Twenty years after experimenting with and making big investments in online news, newspaper publishers have yet to develop a viable business model to support online journalism, nor have they created a product that's competitive in relation to the print counterpart. According to Chyi, "Not only does the audience respond to the print edition more favorably, advertisers vote for the print edition with their ad dollars ... The dead-tree edition still account[s] for as much as 85% of total ad dollars since the recession, ... [and] for every $16 in print advertising revenue lost, only $1 in digital ad dollars was gained" (Chyi 2013, pp. 9, 57, 64).[8]

Unable to recover lost print ad revenues, publishers have tried various types of digital subscription models. In 2014, more than 450 newspapers charged for online content (Chyi 2013, p. 86). A recent American Press Institute survey found that 77 of 98 papers with circulation over 50,000 require readers to pay for unlimited online access. Of the 77, 71 launched subscription plans within the past five years, although the potential revenues generated are "still murky at best" (American Press Institute 2016, pp. 1-4).

Summing up the problem with online newspapers, Chyi argues that "given information surplus, news commoditization, and the perceived inferiority of online content, charging anything for general interest news in digital format has proven difficult. But many papers are still doing it..."(Chyi 2013, p. 87).[9]

Despite the lack of consensus on a viable business model to sustain digital journalism, there has been significant growth in online-only journalism and a proliferation of digital-only outlets in recent years.

A 2014 Pew Research Center report attempted to "put a number on the shifting journalism landscape"[10] to account for editorial staffing at 30 major digital news organizations and 438 smaller ones. Those 468 outlets, the vast majority of which started in the past decade, have produced almost 5,000 full-time editorial jobs. Overall, there has been a relatively quick shift of talent to digital news jobs. Some of these startups have grown to maturity and earn a profit, while many continue to lose money (Mitchell and Page, March 2016, pp. 2-3).

In 2015, advertisers spent almost $60 billion on digital ads, up 20% from 2014 and from $32 billion in 2011. Digital accounts for one-third of all ad spending on any platform, while mobile consists of over half of the $31.6 billion spent on digital. (As recently as 2010, mobile comprised less than 1% of total digital ad spending.) This is no surprise given that mobile audiences are driving the growth in digital media, especially at the digital natives. Digital is second behind only television as the source from where many Americans often get their news. The main beneficiaries, however, continue to be social media and technology companies, especially Google and Facebook which, along Microsoft, AOL, and Yahoo, generated $30.9 billion out of a total of $50.7 billion in domestic ad revenues (Mitchell and Page, April 2015, pp. 17-21; Mitchell et al., June 2016, pp. 45-48, 52).

Despite the recent flow of investment dollars into digital natives,[11] long-term viability is still not assured. This doubt is warranted by the difficulty that legacy news organizations have had in producing enough revenue, particularly advertising dollars (Jurkowitz, 3/26/14).

In the area of employment growth, the largest 30 outlets have been responsible for creating over 3,000 jobs or 102 positions per outlet. The six largest employers are *VICE* (1,100),[12] *Huffington Post* (575), *Politico* (186), *BuzzFeed* (170), *Bleacher Report* (140), and *Gawker* (132). Five of the 30 on the list have become unionized since this report was issued. The other 438 organizations surveyed produced only 4.4 jobs per outlet. Some of these positions were created circa 2005, with many after 2010. A number of these organizations have filled voids left by legacy media in the areas of local news, international coverage and investigative journalism (Mitchell and Page, March 2014, pp. 10-15).

The top five digital-native news entities[13] in terms of the total number unique visitors (as of January 2015) are *HuffingtonPost.com* (100 million), *BuzzFeed.com* (77.9 million), *BleacherReport.com* (44.2 million), *BusinessInsider.com* (37.3 million), and *CNET.com* (30.5 million). Newly unionized outlets in the top ten include *HuffingtonPost.com*, *VICE.com* (15.6 million) and *Gawker.com* (14.1 million).

Two recent trends in digital-only media have the potential to affect union formation and collective bargaining—concentration and consolidation and an uncertain business model. The three biggest online publishers and news aggregators provide 20% of all news consumed online, while the top ten publishers, which together own 60 news sites, accounted for 47% of total online traffic to news content in 2015. The next largest 140 accounted for most of the other half (Sass, 5/12/16).

Across the media landscape, which includes digital-only, a few large corporations—such as Comcast, AT&T, Verizon, and Charter[14]—and newer outlets like Vox Media, which recently acquired *ReCode*,[15] there is a trend toward consolidation, scale, and vertical integration in order to earn enough ad revenue to be profitable. This is increasingly becoming more difficult when the duopoly of Google and Facebook reap more than 40% of the digital ad market in the U.S. In the first quarter of 2016, they captured 85 cents of every *new* dollar spent on online advertising (Doctor, 5/28/15 and 4/13/16; Alpert, 8/19/15; Eavis, 4/23/16).

Digital-only media also is concentrating geographically on both coasts, especially in New York City. Smaller hubs are taking shape in the Bay Area, Los Angeles, Miami, and Washington, D.C. The upshot is that journalism jobs are migrating to where the workforce will likely be more politically, socially, and culturally liberal. Journalist Michael Massing refers to the growing geographical concentration in lower Manhattan as "forming a sort of journalistic counterpart to Silicon Valley and replicating the parochialism of the New York media elite" (Massing, 6/25/15; Benton, 3/25/16).

While geographic clustering and industry concentration and consolidation have the potential to make labor organizing easier, recent industry financial troubles could slow or reverse union gains. In April 2016, *BuzzFeed* cut its revenue goal in half, setting off concern that digital media was bumping up against challenging headwinds (*Recode*, 4/13/16). The evidence for this isn't hard to find. According to *BuzzFeed*: "There's blood in the pixels again. Since the beginning of last year, media companies with large U.S. digital operations have been shedding employees and even shutting down entirely (e.g., *Gigaom* and *Al Jazeera America*). In total, more than 1,000 job cuts have been announced over the last 12 months, and industry watchers fear more to come" (Zeitlin, 4/12/16).

The problems are rooted in the increasingly more treacherous web advertising business, the plateauing of website traffic at many large sites, and challenges from automated advertising and ad-blocking tools, which have led to estimated global industry revenue losses of over $20 billion, and an oversupply of online advertising impressions that generate pennies on the dollar in revenue and are often ignored.[16] Perhaps more fundamental is the transition from an Internet of websites to an Internet of mobile apps and social platforms, controlled by Facebook, Google, and Twitter (Herrman, 4/17/16; Willens, 4/27/16; Yang, 5/3/16). These headwinds, according to *Mother Jones*, could mean a slow popping of the digital media bubble or—as others contend—a potential strategic shift to TV, which earns substantial total revenue (Alpert, 10/12/15; Hagey, 11/2/15; Bauerlein and Jeffery, 4/28/16). It is in the context of recent rapid growth in an uncertain business environment that union formation has been taking shape.

A Tale of Two Guilds: The NewsGuild and the Writers Guild of America-East

Two unions have been organizing digital newsroom employees—the NewsGuild (NG) and the Writers Guild of America, East (WGAE). The NG is the largest and oldest of the two—sometimes competing—unions. It was founded as the American Newspaper Guild in 1933 and absorbed by the Communications Workers of America in 1997. It currently has about 25,000 members in the United States and Canada, down from a peak of about 32,000 in 1986. Roughly 2,000 members work in digital media, including editorial employees at a few digital-only newsrooms (Richardson, 1/28/15; Alpert, 8/19/15; NewsGuild website, www.newsguild.org/).

In September 2015, the NG secured a $500,000 Strategic Industry Fund grant from CWA to organize digital newsrooms and hired a handful of organizers in NYC and D.C. The NG believes that this initiative reflects a core principle: "that quality journalism isn't sustainable without quality jobs. And democracy isn't sustainable without quality journalism" (*The Guild Reporter*, Fall 2015, pp. 2, 6). Moreover, NG president Bernie Lunzer believes that the union is the right one for digital-only employees because it has a long history of representing journalists in a bottom-up manner, is "platform neutral," and has devoted half a billion dollars to organizing. The time also is right because many startups have stabilized and are profitable (*The Guild Reporter*, Fall 2015).

The Writers Guild of America, East (WGAE) represents almost 4,000 workers, including writers for TV, film, documentaries, and broadcast and radio news. The WGAE has a West Coast counterpart and, like the NG, is affiliated with the AFL-CIO. Executive Director Lowell Peterson has been central to organizing activities in digital media. Despite its smaller size, the WGAE has been relatively more successful organizing digital editorial workers. It currently represents employees at *Gawker*, *Salon*, *VICE*, *ThinkProgress*, and *The Huffington Post*, the largest unit to date. These victories between June and October 2015 are returns to the WGAE's 2009 organizing initiative, "Writers Guild 2.0" (WGAE website; Warren, 8/12/15; Richardson, 9/16/15).

A third, more inchoate organization, Media Workers Unite (MWU), was established by journalist and activist Mike Elk in the summer of 2015. Until 2015 Elk was a labor reporter at *Politico* but was discharged allegedly for attempting to organize the newsroom. Elk's father, aunt, and uncle were organizers for the United Electrical, Radio, and Machine Workers of America (UE), a radical union. In October 2015, the MWU held its inaugural convention in Louisville, K.Y. The MWU's 12-point plank covers issues such as overtime, diversity, ethics, employment discrimination, training and technology, freelancer and intern rights, and union elections free from interference from management. To date, the MNU has not conducted any formal organizing drives (Beaujon, 8/13/15; Mullin, 8/20/15; (mediaworkersunite.com/).

"Sure the Snack Bar Is Great, But a Union Contract Would Be Even Sweeter!": Issues Driving the Demand for Unions[17]

In the same way that technology has "disrupted" legacy media, union organizing has upset the industry's business model, which is undergirded by a workforce that is young (20s and 30s), highly educated, skilled, but low-paid and insecure. Many of these journalists were once content just to have a job in the aftermath of the Great Recession. No longer. Evoking the first wave of newsroom unionism, *The Nation's* Michelle Chen reminds us that "newspaper and entertainment workers helped rebrand industrial unionization during the 1930s, and redefined the meaning of the industry itself, as writers, actors and other creative workers mobilized to prove the artistic nature of their craft work did not preclude the need for rights and standards at work" (Chen, 10/5/15). The current wave of newsroom journalism holds out hopes for defining professional and labor standards in digital media and perhaps for a revived labor movement.

There appear to be similarities, however, between then and now. Despite their upper-class backgrounds, "a lot of (young digital) journalists share the same basic concerns as factory workers, or home care workers, or other types of workers who traditionally turn to labor unions in this country," according to *Huffington Post* labor reporter Dave Jamieson" (Moses, 8/13/15; Cohen, 9/1/15).

Younger workers might be less likely to be union members, but they hold more favorable views of unions than do other age groups, according to a Pew Research Center poll. For example, 55% of Americans between 18 and 29 years old said they viewed unions favorably compared with 29% who didn't. For older respondents, the number dropped to 46% favorable and between 40 to 43% unfavorable (Pew Research Center, 4/27/15). Similarly, a recent Gallup poll found that while 58% of Americans approved of unions and 37% want them to have more influence, 66% of adults between 18 and 34 years old approved of unions and 44% want unions to have more influence. Approval ratings for the 18- to 34-year-old group are the highest of any demographic category (Gallup, 8/17/15).

The motivations for union representation vary by organization, but a set of common factors appear across news outlets. In addition to economics—pay and benefits—working conditions are a major concern. Many journalists, for example, cite stress owing to overwork and the relentless nature of their jobs in a 24/7 news cycle. Mike Elk contends, "Reporters are forced to work long hours when a big news story breaks, but how do you get overtime pay or ask for time off without a union contract?" (Mullin, 8/20/15).

The desire for workers to have a voice in shaping workplace rules also is a top priority. Hamilton Nolan, journalist and lead organizer at *Gawker*, the site of the first successful digital-only drive, explains why the editorial staff unionized. Although "Gawker Media is a very good place to work," he contends that every workplace could use a union, which is the "only real mechanism that exists to represent the interests of employees … This is useful in good times (which our company enjoys now), and even more in bad times (which will inevitably come)" (Nolan, 4/16/15).

At *Gawker*—and elsewhere—employees have expressed concerns about the lack of transparency and inconsistency in work rules and employment policies. Recognizing that the company is well run, pays competitive salaries, and treats its employees reasonably well, Nolan points to certain issues that employees would like to see addressed—pay commensurate with experience and type of work done and a way "to ensure that things like pay and raises are set in a fair, transparent, and unbiased way." To accomplish these and other goals, Nolan calls for "some basic mechanism for giving employees a voice in the decisions that affect all of us here" (Nolan, 4/16/15).

In his missive, Nolan tries to assuage the concerns of management and his co-workers who might oppose or be skeptical of unions: "Nobody is seeking to hurt this company, or plunder it for all it's worth, or find a way to attack people that run it. We're just trying to make it a bit more functional and a bit more fair" (Nolan, 4/16/15).

As the industry and workforce mature, employees also seek sustainable careers, especially since legacy media companies have shed so many jobs and cut short once-stable careers (Maharidge, 3/2/16). One politics reporter at *ThinkProgress* notes that many of the young newsroom employees started their careers in the depths of the recession and had been told by others that they were lucky to have any job. But now she and her colleagues have learned "you can't eat exposure, you can't pay your rent with experience, as valuable as those things are" (Somaiya, 9/23/15). Gabriel Arana, former senior media editor at *The Huffington Post*, who was involved in the union drive there, echoed a similar sentiment: "A lot of these new-media companies feel like tech companies. But at a certain point, having free snacks at work means less than having a retirement account or a decent salary that you can raise a family on. Digital media is maturing. People want the stability to be able to make a career out of it" (Maharidge, 3/2/16).

How Progressive Are the Media Liberals?: "Words Are Cheap. Unions Can Cost Money"

One of the more impressive aspects of labor organizing in digital-only newsrooms has been the relative ease by which unions formed at *Gawker*, *Salon*, *Guardian US*, *VICE*, *ThinkProgress*, *Al Jazeera America*, and *The Huffington Post.* All but one resulted from voluntary card-check procedures. At *Al Jazeera America*, management demanded an NLRB-supervised election, which the union easily won. This quick burst of successful union organizing is rare U.S. labor relations.

Soon after *Gawker* unionized, Nolan challenged executives at similarly liberal media outlets to respect employees' desires for union representation in a piece entitled, "How Progressive Are the Media Liberals?" Disappointed that executives at the left-leaning publication *Salon* hired a public relations firm and a labor lawyer and acted "markedly different from what progressive values would have dictated" during an organizing drive, Nolan dared them to uphold their political conviction: "Words are cheap. Unions can cost money. This is where we learn how real all of these nice progressive media bosses really are. If they won't support their own workers' union, they were lying the whole time" (Nolan, 7/31/15).

As the discussion below reveals, the role of management is especially important in the organizing process and might suggest ways that U.S. labor policy could be altered to enable workers to express their preferences for workplace representation free from management interference. Research shows that the desire for workplace representation is significant. For example, Freeman and Rogers (1999) found that 32% of nonunion private-sector workers (44% overall) would vote for a union at their workplace, and that larger numbers of workers want some type of representation. In subsequent reports, Freeman notes that this number rose to 53% in 2005 (Freeman 2007a, pp. 26, 35-36; 2007b, pp. 1-2). With private-sector union density rates currently at 6.7% (Bureau of Labor Statistics, 1/28/16), there clearly appears to be a "representation gap" owing to a deteriorating legal and political environment (Farber 2015), the role of management resistance and unfair labor practices during the organizing process (Ferguson 2008), and union decentralization (Fiorito and Jarley 2012).[18]

Organizing the "Quick and Painless" Way: From *Gawker* to *The Huffington Post*

Over a period of six months, from June 2015 to January 2016, editorial employees at seven properties formed unions the "quick and painless way." The 500-plus union-represented employees pale in comparison to the tens of thousands of jobs lost in newspapers and other media, but they have the potential to establish a beachhead from which to increase union density in the future.

Gawker

Backed by the WGAE, the editorial staff at the gossip and media news site initiated an organizing drive in April 2015 that had widespread though not universal support. By late May, the WGAE and Gawker Media, *Gawker's* parent company, headed by co-founder Nick Denton, jointly announced a "first-of-its-kind election whereby the members of Gawker Media's editorial staff will determine whether they wish to form a union." The route to unionization also was atypical: "By mutual agreement we are bypassing [the NLRB] in favor of holding our own secret-ballot vote …[W]e believe the cumbersome and often factious process of unionization is premised on an assumption of complete antagonism between labor and management" (Craggs, 5/27/15; Mullin, 5/28/15).

Veteran labor reporter Steven Greenhouse opined on the significance of the upcoming vote: "If the unionization effort succeeds, it will be a big PR boost for the ailing labor movement. It will show that unions, which have focused in recent years on organizing low-wage workers, can also attract hip, highly educated workers, many of them Ivy League graduates" (Greenhouse, 6/2/15). It did succeed. *Gawker* staffers voted 80-27 for the WGAE (Richardson, 6/10/15). Denton, who was "intensely relaxed" about the prospect of a unionized workroom said, "There's no reason that so many U.S. workplaces are contentious, and I'm very pleased *Gawker* is leading the movement in the online media world toward collaboration and inclusion" (Warren, 6/4/15; Meyerson, 6/9/15).

In addition to be the first to organize, Gawker also was the first digital-only newsroom to reach a contract settlement, in February 2016, that was ratified overwhelmingly by 99 editorial employees in early March. The seven-page contract, which runs for three years, establishes minimum salaries ranging from $50,000 for staff reporter to $150,000 for editor-in-chief. It also provides annual raises, health and retirement benefits, paid time off, a grievance procedure, severance pay, editorial independence, limits to contractor labor, a diversity committee, monthly meetings between the union and management, and union security. The agreement also gives management operating flexibility and includes an at-will clause, which is uncommon in most labor contracts but common in WGAE ones (Richardson, 3/9/16; *Collective Bargaining Agreement, Writers Guild of America, East and Gawker Media*, effective 3/1/16-2/28/19).

Salon

In early July 2015, following the *Gawker* election, all 26 editorial staffers at Salon Media sought WGAE representation. The San Francisco-based outlet, which features politics, breaking news, and technology, was founded in 1995. It employs half its editorial staff in NYC with the remainder in San Francisco and others working remotely across the country. In a public letter, employees stated that a union would provide an "opportunity to establish standards and practices in Internet journalism" (Jamieson, 7/2/15). Moreover, it claimed, employees "are doing this because we believe in our publication and want it to be successful … We believe this organizing campaign is a positive and public way for us to put those [progressive] values into practice, right here at home" (Richardson, 7/8/15).

Salon's editor-in-chief David Daley was supportive of employees' wishes: From its very inception, [*Salon*] proudly embraced progressive values and a commitment to our workers and to labor" (Jamieson, 7/2/15). Despite his favorable remarks and employees' unanimous union support, it took a month of foot-dragging before management recognized the WGAE (Richardson, 8/5/15).

Guardian US

In late July 2015, all 45 U.S. newsroom employees of the British news organization's rapidly growing U.S.-based operations cast their vote for the NewsGuild in a private American Arbitration Association-supervised election.

President Lunzer signified the victory this way: "Digital media is growing up, and it's time our digital reporters received the same benefits and protections as their print media colleagues." A company spokesperson remarked, "We are happy to voluntarily recognize the [NewsGuild] and look forward to working constructively, in best Guardian tradition, with the *Guardian US* editorial staff who have voted in favor of collective representation" (Alpert, 7/29/15). The U.S.-based staffers joined the company's tradition of union representation in the UK and Australia (Bhatia, 7/29/15).

VICE

Originally founded as a punk rock print magazine called *VICE* in Montreal in 1994, VICE Media has roughly 700 U.S. employees in its Brooklyn office, about 10% who are journalists. On August 7, *VICE* agreed to voluntarily recognize the WGAE through card-check procedures. With 75 editorial employees in the unit, *VICE* became the largest digital media company to unionize at the time. It was also well-heeled, being valued at $2.5 billion in 2014 (Alpert, 8/7/15).

In a letter to co-founder and CEO Shane Smith, employees said they "are proud of the work we do here at *VICE*. We love being part of a company that is changing media and having an impact on the world … We believe that a union is a logical next step for the long-term legacy of the company." The young employees' main concern was having a voice in management decision making (Richardson, 8/12/15).[19]

On April 21, 2016, following the lead at *Gawker*, *VICE* employees ratified a three-year labor agreement by a 73-2 vote. The contract includes minimum salaries of $45,000/year, raises totaling 29%, protections of the current health plan, permission for unit employees to engage in non-*VICE* work, compensatory time off for work on weekends or other scheduled days off, guaranteed severance pay, and a commitment to meet on a monthly basis to discuss work-related issues. Like the *Gawker* contract, this one also includes an at-will clause. Chief Operating Officer Alyssa Mastromonaco seemed pleased with the pioneering contract: "This is great news: Both sides have come together and struck a deal that continues to propel *VICE* forward as media's most innovative and entrepreneurial workplace" (WGAE, 4/15/16 and 4/21/16; Rose, 4/20/16).

Al Jazeera America

In late summer 2015 editorial employees at the Qatari government-owned news outlet moved toward unionizing. In its mission statement, employees said they "want to work with *Al Jazeera* management in a spirit of honesty, integrity, and transparency—both to produce the news and to improve the workplace in which we produce it." Employees were upset at the "lack of transparency, inconsistent management and a lack of clear redress" at the digital outlet. Specifically, they were displeased with discrepancies in pay, confusion about job assignments, the evaluation process, and workplace climate (Calderone, 9/3/15; Richardson, 9/9/15).

Al Jazeera America (*AJAM*) staffers chose the 2,800-member NewsGuild of New York to represent them. Failing to achieve voluntary recognition, the union filed for an NLRB election for the 50-member bargaining unit in late September. Despite management's protest of a few unit positions, the union won in a lopsided 32-5 vote. Management proclaimed that it looked "forward to concluding a collective bargaining agreement and will continue to work closely with all of our staff to make working at *AJAM* a positive experience" (Warren, 9/29/15 and 10/6/15; Richardson, 10/14/15).

The union victory at *AJAM* proved to both challenging—it has been the only contested NLRB election so far—and pyrrhic because the company announced it would shutter its digital and cable news operations by the end of February 2016 because of a drop in energy prices that sustained this government-owned business (Hagey and Flint, 1/13/16; Sutton, 2/8/16). Even without a contract, the NewsGuild negotiated a severance package stronger than the company's initial offer and won guarantees that employees would receive the benefits even if they secured a new job before *AJAM* closed down for good (Hartman, 3/7/16).

ThinkProgress

Editorial employees at the left-leaning public affairs website *ThinkProgress* joined the WGAE and won voluntary recognition in late September 2015 from its parent, the Center for American Progress (CAP). CAP is a progressive D.C.-based public policy research and advocacy organization founded in 2005 by lawyer Jedd Legum. Like editorial workers at other digital-only properties, the *ThinkProgress* staff desired to improve their pay and benefits, have voice in the workplace, and create a more structured workplace (Richardson, 9/30/15).

Support for the union by the roughly 30 editorial employees was unanimous. In a letter to CAP president Neera Tanden, staffers noted that the website and CAP have "long publicized the many benefits of unionization for individual workers, the middle and working class, and the economy as a whole. Here at *ThinkProgress*, we believe having a union will help us protect and elevate the good conditions we currently enjoy in our workplace, both for ourselves and for all who come after us" (Jamieson, 9/25/15). Legum agreed: "We support and recognize the move by *ThinkProgress* staffers to join the (WGAE). Our shared goal is to make *ThinkProgress'* exceptional team of writers and editors even stronger and continue our cutting-edge work covering the issues that matter most" (Richardson, 9/30/15).

The Huffington Post

In 2011, AOL purchased *The Huffington Post* for $315 million. In June 2015, AOL and *The Huffington Post* were acquired by the telecom giant Verizon for $4.4 billion. A few months later, editorial employees at this liberal-left news site, founded by Arianna Huffington in 2005, formed an organizing committee, assisted by the WGAE, to help them achieve workplace voice, editorial freedom, management transparency, better pay, clearer job duties, a more efficient hiring process, and more diversity in its workforce. If successful it would become the largest digital-only bargaining unit (Kaufman, 6/30/15; Ember, 10/7/15; Richardson, 12/2/2015).

In late 2015, employees asked management to voluntarily recognize the union given that 220 of the roughly 350 eligible employees had signed authorization cards. A few months prior, Huffington publicly declared her support for the union. Finally, in January 2016, employees voted "overwhelmingly" for representation by the WGAE in a card-check vote that created a 262-person unit. The WGAE's Lowell Peterson remarked on the victory: "*The Huffington Post* establishes a critical mass of support for collective bargaining in digital media ... This movement empowers the people who write, edit, and produce content to build sustainable careers doing the work they care about." Arianna Huffington was equally optimistic: "We look forward to a great relationship with our union as we work together to take *The Huffington Post* to new heights across the globe" (Richardson, 1/20/16).

Since the *Huffington Post* vote, no other digital-only media properties have been unionized, and only two of the six remaining newsrooms have secured labor agreements. At present, close to 500 editorial employees are represented by unions, but both the WGAE and the NG remain committed to organizing new units. Although both print and digital journalism continue to face challenging economic circumstances, plenty of opportunities exist to create collective bargaining relationships for tens of thousands of media workers.

Conclusions and Implications

This paper has explored the recent wave of successful labor organizing in digital-only media, a relatively new but growing media segment. This burst of union activity occurred over a relatively short period of time (six months) and without much management opposition, both anomalous in U.S. labor relations.

Unlike the older, more seasoned, union veterans in legacy media, digital-only media workers are younger, more educated and affluent, and have less experience with labor unions. Still, surveys reveal that younger workers, especially more liberal ones, support unions in greater numbers than other age groups (Pew research Center, 4/27/15; Gallup, 8/17/15). Moreover, as Budd (2010) uncovered using National Longitudinal Survey of Youth data, while most workers are likely to be unionized in their forties, many experience unionism for the first time as younger workers. He found that almost 65 percent of the entire cohort studied was unionized in at least one of their jobs by age 40/41, with half first unionized by 23 and three-quarters by 27. One implication, Budd notes, is that "U.S. labor unions have an important, and probably overlooked the opportunity to develop a supportive, firsthand relationship with quite a large fraction of the U.S. workforce" (Budd 2010, p. 222). In the case of digital-only news workers, unions are, in fact, taking advantage of this opportunity that many other unions have not yet capitalized on. As Table 1 shows, in the seven successful organizing drives to date, editorial employees voted overwhelmingly—in some cases unanimously—for union representation.

The mostly young digital-only workforce is not only more educated and likely more liberal than older workers, but so too are their employers, which has helped with labor organizing. Instead of fighting employees' attempts to unionize the way most American employers do, in six of the seven cases, these digital-only employers voluntarily recognized unions via some form of card check. Even at *Al Jazeera America*, where employees voted in an NLRB-supervised election, management opposition was tempered. While most employers do not hold progressive beliefs, and not all progressive employers support unions in their workplaces, the seven cases discussed above suggest that amending labor law to enable workers to express their desires for a union—free from management interference—may restore the original promise of the National Labor Relations Act. As Freeman and his colleagues (1999, 2007a, 2007b) have found, there is a large "representation gap" in the U.S. that is also revealed in the case of digital-only media. In addition, as this study and prior research make clear, card-check procedures are an effective way to close this gap (Eaton and Kriesky 2001, 2009).

Having established a foothold in digital media, both in the online operations of traditional newspapers and the digital natives, both the NewsGuild and the Writers Guild are committed to growing their memberships and increasing union density in new media in order to establish industrywide professional and labor standards that will benefit their members and uphold the nation's democratic principles, which gave the news industry its revered status as the Fourth Estate.

This effort, however, remains fraught by the unsettled nature of the industry and other events. For example, in addition to industry challenges discussed above, a $140 million legal judgment against Gawker Media Group, in a case involving the former wrestler Hulk Hogan, forced it to file for bankruptcy and put itself up for sale. The fate of both Gawker Media and the first collective bargaining agreement in a digital-only newsroom—and perhaps in others—will have to await the public auction of the company in late July 2016 (Ember, 6/10/16; Richardson, 6/15/16).

Even without collective bargaining, digital-only employers could still address employees' concerns over compensation, voice, transparency, fairness, stable careers, and other issues that have become salient to their mostly millennial workforce. These concerns are also shared by many workers in both the "gig economy" and the low-wage service sector. With industrial and craft union labor in decline for decades, the future of the American labor appears to reside in the fortunes of these disparate groups of workers who have been agitating for a stronger voice in the workplace to help them achieve better pay and benefits, higher quality jobs, employment stability, and personal dignity. Whether or not the recent actions by fast-food, retail, and media workers will reignite the American labor movement remains to be seen, but media workers are well-positioned to publicly chronicle these struggles in the same their counterparts in smoky and noisy newsrooms did in the 1930s.

Endnotes

[1] New media has been referred to by other names that will be used interchangeably here: digital natives, digital-only media, online journalism, etc.

[2] In January 2015, the Newspaper Guild-CWA officially removed "paper" from its name. The new, NewsGuild-CWA, reflects changes in media that have occurred over the past few decades and the union's broader jurisdiction, which includes broadcast news and nonprofits (Richardson, 1/28/15).

[3] The NewsGuild represents journalists at the digital-only *Daily Beast*, which it inherited through a merger with the unionized *Newsweek* in 2011. After owner IAC sold *Newsweek*, the Guild negotiated a separate contract with *Daily Beast* (DePillis, 1/30/15). The NewsGuild also represents editorial employees at politically left websites *Truthout* and *In These Times*, print magazine with a web presence (Sterne, 9/1/15).

[4] The Newspaper Association of America, the main industry trade group, ceased reporting data after 2014.

[5] Estimates are based on data from the remaining seven publicly traded newspaper companies, which own about a quarter of U.S. dailies (Mitchell et al., June 2016, p. 9).

[6] Average weekday circulation fell 7% between 2014 and 2015, the largest decline since 2010. The decline was entire due to print circulation, which fell 9%, while digital circulation rose 2% (Mitchell et al., June 1016, p. 10).

[7] Despite the doom and gloom predictions by many media analysts and industry insiders, media economist Marc Edge argues that metro dailies in both the U.S. and Canada have been adaptable, viable, and profitable. Though

revenues, earnings, and profit levels declined, none of the companies suffered an annual loss for its newspaper division, according to financial reports for 11 publicly traded U.S. newspaper companies and five in Canada between 2006 and 2013. The majority of companies emerged from the Great Recession with double-digit profits. The biggest problem for most of these companies was burdensome debt accrued for purposes of acquisitions for which some companies declared bankruptcy (Edge 2014a).

[8] In a study of three cities, the Pew Research Center finds that almost half of newspaper readers rely only on the print edition, which remains an important part of newspapers' business model and garners over three-quarters of their advertising revenue (Barthel, 1/6/16).

[9] Marc Edge, in *Greatly Exaggerated: The Myth of the Death of Newspapers* (2014b) also refutes the conventional wisdom that the print newspaper is dead and that the future lies in online journalism. Edge, however, provides a more comprehensive and historical account than Chyi.

[10] The Pew Research Center's annual *State of the News Media 2015* reaffirms the challenges of identifying the nature and scope of the industry: "Overall, digital news entrants and experimentation, whether from longtime providers or new, are on the one hand now so numerous and varied that they are difficult to keep track of. On the other hand, the pace of technological evolution and the multiplicity of choice—from platforms to devices to pathways—show no signs of slowing down" (Mitchell and Page, April 2015, p. 8).

[11] *VICE* Media has raised $70 million, Vox Media has secured $60 million, and BuzzFeed has received $46 million. Facebook, the largest publishing platform, has $13 billion in cash, $7.9 billion in sales and a 37% operating margin. Investors favor exponential growth in audience size and page views over traditional metrics as revenues and profitability, according to industry analyst Alan Mutter. Investors hope that these new digital ventures will merit an IPO or be acquired by a well-heeled player (Mutter, 4/2/14).

[12] This number includes all full-time employees, not just editorial staff.

[13] Only *The Huffington Post* (4) and *BuzzFeed* (7) are on the list of the ten largest digital news entities, which include all media companies with an online presence.

[14] Charter Communications completed its acquisition of Time Warner Cable in mid-May 2016 and now stands as the second-largest broadband provider in the country behind Comcast (Steele and McKinnon, 5/18/16).

[15] Vox Media and *BuzzFeed* both benefitted from NBC Universal's investment of $200 million each (Doctor, 4/13/16).

[16] Cost per Thousand, or CPM, is the standard measurement for online ads. The industry might be better off changing the metric to some measure of "time spent," which would enable publishers to charge advertisers more (Willens, 4/27/16).

[17] Mike Elk to fellow journalists at *Politico*, where he unsuccessfully tried to agitate co-workers to organize a union there (Wemple, 7/2/15).

[18] The literature on union organizing is too voluminous to address here. The articles cited above contain many referenced to the most important works related to organizing.

[19] *VICE UK* staffers, inspired by union activity in the U.S., commenced an organizing drive under the auspices of the National Union of Journalists in February 2016. Employees have sought a fair and transparent compensation system, job security, work-life balance, career opportunities, and to relieve the ongoing stresses and strains of the never-ending news cycle (Sweney, 2/8/16). In late Spring 2016, 130 employees from different work areas at *VICE* Canada selected the Canadian Media Guild to assist them in achieving "some basic standards," more consistency and transparency in work rules and pay, improved benefits, just cause protections, and clear journalistic standards (*The Guild Reporter*, Summer 2016, p. 1).

Select Bibliography

Bibliographic note: Given the length of the original bibliography, this one is more selective. It includes all the academic sources and the main industry reports. Other key sources used here include, but are not limited to, Pew Research Center, Poynter Institute, Nieman Journalism Lab, *Reflections of a Newsosaur* blog, Bloomberg BNA's *Labor Relations Week* and *Daily Labor Report*, *The New York Times*, *The Wall Street Journal*, and many of the same digital-only media outlets covered in this paper. A full bibliography is available from the author stangerh@canisius.edu.

Chyi, H. Iris. 2013.Trial and Error: U.S. Newspapers' Digital Struggles Toward Inferiority. Pamplona, Spain: Universidad de Navarra.

Eaton, Adrienne E., and Kriesky, Jill. 2001. "Union Organizing Under Neutrality and Card Check Agreements." *Industrial and Labor Relations Review*, Vol. 55, no. 1 (October), pp. 42-59.

Eaton, Adrienne E., and Kriesky, Jill. 2009 (January). "NLRB Elections Versus Card Check Campaigns: Results of a Worker Survey." *Industrial and Labor Relations Review*, Vol. 62, no. 2, pp. 157-172.

Edge, Mark. 2014.a "Newspapers' Annual Reports Show Chains Profitable." *Newspaper Research Journal*, Vol. 35, no. 4 (Fall), pp. 66-82.

Edge, Mark. 2014b. *Greatly Exaggerated: The Myth of the Death of Newspapers.* Vancouver: New Star Books.

Farber, Henry. 2015. "Union Organizing Decisions in a Deteriorating Environment: The Composition of Representation Elections and the Decline in Turnout." *Industrial and Labor Relations Review*, Vol. 68, no. 5 (October), pp. 1126-1156.

Ferguson, John-Paul. 2008. "The Eyes of the Needles: A Sequential Model of Union Organizing Drives, 1999-2004." *Industrial and Labor Relations Review*, Vol. 62, no. 1 (October), pp. 3-21.

Fiorito, Jack, and Jarley, Paul. 2012. "Union Organizing and Membership Growth: Why Don't They Organize?" *Journal of Labor Research*, Vol. 33, no. 4 (December), pp. 461-486.

Freeman, Richard B. 2007a. "Can the United States Clear the Market for Representation and Participation?" In Freeman, R. B., Boxall, P., and Haynes, P., *What Workers Say: Employee Voice in the Anglo-American Workplace.* Ithaca: Cornell University Press.

Freeman, Richard B. 2007b (February 22). "Do Workers Still Want Unions? More Than Ever." (Briefing Paper #182). Economic Policy Institute. www.sharedprosperity.org/bp182.html

Freeman, Richard B., and Rogers, Joel. 1999. *What Workers Want.* Ithaca: Cornell University Press.

Mitchell, Amy. Project for Excellence in Journalism (PEJ), Pew Research Center. 2014 (March 26). "State of the News Media 2014: The Growth in Digital Reporting: What It Means for Journalism and News Consumers." www.journalism.org/files/2014/03/Shifts-in-Reporting_For-uploading.pdf

Mitchell, Amy, Jesse Holcomb, and Rachel Weisel. Pew Research Center. 2016. (June). "State of the News Media 2016." www.journalism.org/2016/06/15/state-of-the-news-media-2016

Mitchell, Amy, and Dana Page. Project for Excellence in Journalism (PEJ), Pew Research Center. 2015 (April 29). "State of the News Media." www.journalism.org/files/2015/04/FINAL-STATE-OF-THE-NEWS-MEDIA1.pdf

XI. LERA Best Papers II

Do Different Stakeholders Matter Differently to Workers? Developing and Applying a Stakeholder CSR Measure to Organizational Behavior Outcomes

Brandon C. Grant
University of Illinois at Urbana-Champaign

Joel E. Cutcher-Gershenfeld
Brandeis University

Robert A. Bruno
University of Illinois at Urbana-Champaign

Corporate social responsibility, or CSR, has traditionally been measured and studied under several names, all of which attempt to measure CSR as a sum of an organization's internal behaviors by compartmentalizing those behaviors by *domain*. The measure developed in this paper focuses instead on the stakeholder *targets* of those behaviors, and in doing so allows researchers to explore target-specific outcomes of CSR perceptions more completely. Additionally, while previous studies of CSR have focused on measurement at the management level, this study measures CSR as perception at the worker level. Study 1 confirmatory factor analysis supported four-factor stakeholder-focused structure, which fit the data significantly better than the original four-factor domain structure. Study 2 further develops this stakeholder measure of corporate social responsibility by reducing the scale to three dimensions. Structural equation modeling demonstrates predictive validity for the dimensions on key organizational outcomes such as counterproductive work behaviors and organizational citizenship behaviors.

The past five decades have seen a sizeable increase in scholarly interest in the study of corporate social responsibility (CSR), specifically with respect to the role of stakeholders to an organization's CSR initiatives (i.e., Camplin 2011; Freeman 1984; Rolland and O'Keefe Bazzoni 2009; Werther and Chandler 2005, 2010). Carroll (1979, 1991, 1999) defines CSR as the "social responsibility of business encompass[ing] the economic, legal, ethical, and discretionary expectations that society has of organizations at a given point in time" (Carroll 1979, p. 500). Carroll also provides a useful definition of stakeholders as part of a larger discussion on how stakeholders are inextricably linked to an organization's CSR practices, defining stakeholders as "those groups or persons who have a stake, claim, or an interest in the operations and decisions of the firm" (1991, p. 43). Examples of stakeholders include workers, labor unions, communities, shareholders, the environment, customers, suppliers, and the like. Since Carroll (1979), many scholars have begun to explicitly incorporate the term "stakeholder" in how they define and conceptualize CSR, moving forward what we refer to as *stakeholder CSR*.[1] We define stakeholder CSR as volitional organizational behaviors, actions, and considerations above and beyond financial and legal obligations directed toward any party or who can affect or are affected by said actions, behaviors, and considerations.

Author's address: Brandon C. Grant, School of Labor and Employment Relations, University of Illinois at Urbana-Champaign, 504 East Armory Champaign, IL 61820

In addition to stakeholder CSR, scholars have recently begun to take interest in the relationship between workers' perceptions of their organization's justice-related initiatives—such as CSR—and behavioral outcomes (i.e., Gond, El-Akremi, Igalens, and Swaen 2010), yet when taken together, no known research to date has undertaken the task of developing a measure that captures those worker perceptions. As the two studies in this paper demonstrate, one of the most beneficial uses of such a measure is demonstrating how different stakeholders (specifically, external society) matter differently to workers when considering behavioral outcomes.

For some time, scholars have emphasized the importance of targets of organizational behaviors and how they tend to operate differently and carry with them different antecedents and outcomes (i.e., Bennett and Robinson 1999). Recently scholars have taken an interest in exploring the role of *stakeholder corporate social responsibility (CSR)*, which emphasizes the importance of *targeted stakeholders* in determining leadership behaviors, such as transformational leadership, and subsequently follower outcomes (i.e., Groves and LaRocca 2011; Groves 2014) while other scholars have sought to link CSR to worker outcomes such as *organizational citizenship behaviors* and *counterproductive work behaviors*, as well as trust, commitment, and job satisfaction (Gond, El-Akremi, Igalens, and Swaen 2010).

"The development of [CSR] practices [and the measurement thereof] is associated with the emerging idea that [CSR] is a good market practice … [and] designates the activities and organizational processes adopted by businesses to meet their social responsibilities" (Maignan, Ferrell, and Hult 1999: pp. 455-6). In an effort to understand the relationship between CSR at the organizational level and perceptions of organizations by consumers and society, Maignan et al. (1999) developed a four-factor measure of socially responsible "citizenship" behavior. Motivated by an interest in interpreting the drivers of organizations' active decision to engage in socially responsible behavior, these authors developed a scale that groups activities "representative of corporate citizenship … typifying economic, legal, ethical, and discretionary citizenship toward the three stakeholder groups of interest: employees, customers, and public stakeholders" (p. 460). While the predictive validity of a measurement grouping items by type of activity is strong for employee commitment and customer loyalty, Maignan et al. (1999) are unable to demonstrate the extent to which each targeted stakeholder groups (i.e., stockholders, workers, unions, and society) matters differently from one another in predicting behavioral outcomes. This paper serves to remedy this gap in the literature.

Research in the social sciences has traditionally emphasized the importance of worker attitudes in managing organizations (i.e., Grant 2007; Organ 1988; Penney and Spector 2005), noting that when workers have favorable attitudes toward their jobs and their organizations this can lead to desirable outcomes such as increased levels of organizational citizenship behavior (Organ 1988) and decreased levels of counterproductive work behaviors (Bennett and Robinsons 2001; Penney and Spector 2005). Taking into account the importance of worker perceptions of their organizations, the importance of targets of their organizations behaviors, actions, and decisions (i.e., stakeholder targets), and the significance of targets of worker behavioral outcomes (i.e., the organization vs. other individuals), this paper links these three domains and synthesizes them while simultaneously establishing an important and useful measure of perceived stakeholder CSR.

Taking Freeman's definition of stakeholders, the potential number of a given organization's stakeholders is conceivably limitless, which forces both organizational managers and scholars to carefully select stakeholder groups for consideration. A measure of CSR that examines worker perceptions of their organization's CSR activities directed toward specific stakeholder groups—yet remains sufficiently malleable to be adapted for additional stakeholder groups in the future—is needed. Using a four-factor measure (shareholders, society at large, workers as individuals, and the union), this paper seeks first to establish a model of stakeholder-focused perceptions of CSR, and second to demonstrate the predictive validity of this model for two behavioral outcomes: organizational citizenship behaviors (OCBs) and counterproductive work behaviors (CWBs).

Stakeholder Corporate Social Responsibility and Its Measurement

Corporate social responsibility (CSR) is an important subject of inquiry, yet the study and measurement of CSR are relatively inconsistent and empirical examinations have yielded inconclusive results (McWilliams, Siegel, and Wright 2006). At the heart of this dilemma is the lack of a measurement device that is applied consistently across studies, and is appropriately broad in scope. Building on the work of Aguilera, Rupp, and colleagues, Gond et al. (2010) explicitly connect CSR to perceived social justice by workers, and subsequently connect these perceptions to behavioral outcomes. Individuals internalize perceptions of operations and decision making by asking themselves if a decision made about or an action targeting themselves was fair (i.e., Colquitt 2001), and they similarly consider whether a decision made about or action targeting others (such as fellow workers, their union, or society at large) was fair (Rupp 2011).

The interpretation, definition, and measurement of CSR have become clouded with the emergence of various conceptualizations of CSR such as stakeholder models, ethical models, and social obligation models (see Carroll 1999: p. 290). Ethical models of CSR focus on the moral duty an organization has to do what's right irrespective of whether the moral actions are good for the business and emphasize the humanitarian/moral obligations that organizations should embrace. Social obligation models emerged in the 1950s as some of the first justifications for CSR, which imply that an organization, permitted to exist and operate by society, is indebted to society and is obligated to do right by society. Stakeholder models arrived in the mid-1990s and specified targets and actors (see Maignan and Ferrell 2004 for a summary). This article takes the stakeholder perspective of CSR and demonstrates how organizing measurement items to emphasize specific stakeholder groups not only fits the data better than the four-factor model proposed by Maignan and colleagues (1999) but has predictive power for key behavioral outcomes.

The Measurement of Corporate Social Responsibility

One challenge that researchers studying CSR face is that it has many definitions, which leads to inconsistent measurement and ultimately inconclusive findings. Given the wide scope of actors, stakeholders, and motives that contribute to an organization's CSR initiatives, a definition of CSR that captures this breadth is important to developing a scale that adequately measures CSR across the wide stakeholder landscape, but also across the many organizational levels at which CSR operates (Aguilera and Rupp 2007).

As Wood and Jones point out, while the fit and predictive power of adaptations of Carroll's four-domain model of CSR are inconclusive and limited, they do emphasize that, generally, "the *fundamental* responsibility of business is economic … [and] 'the business of business is business' " (p. 233). In practice, this suggests that the other three domains of CSR (legal, ethical, and discretionary) take a backseat to economic performance/responsibility, providing firms with justification to neglect these other forms of social responsibility in the name of profits. Stakeholders—those who have an effect on or are affected by an organization's actions and behaviors—become linked to the definition and operationalization of CSR. In failing to include stakeholders in the operationalization of CSR, organizations miss out on key opportunities to influence desirable outcomes at the worker level. As the studies in this paper will demonstrate, different stakeholders do in fact matter differently in determining worker-level outcomes.

Scale and Item Development

Hypotheses

The perceived relationship between a worker's union and her employer has a well-documented history of importance to workers, whether positive and cooperative (i.e., Balser and Winkler 2012; Leana, Ahlbrandt, and Murrell 1992), adversarial (i.e., Hammer and Stern 1986; Kelly and Kelly 1991), or inappropriate (i.e., Harcourt and Lam 2007[2]). Other behavioral outcomes of scholarly interest include organizational citizenship behavior (OCB) and counterproductive work behavior (CWB). Generally, one might expect to see a positive relationship between CSR perceptions and OCB as well as a negative relationship between CSR perceptions and CWB. Similar to OCB, counterproductive work behaviors (CWBs) also have been shown to have the same two sub-dimensions (interpersonal and organizational), and tend to operate with similar antecedents (although the literature is less clear than that of OCB; see Dar 2010). With evidence from studies such as Balser and Winkler (2012) indicating that positive, cooperative union-management relations improves worker attitudes toward their employer and Dean (1954) indicating that workers can experience dual loyalty to both their employer and their union, we anticipate that when workers perceive their union to be a target of their organizations' CSR initiatives they will be (a) more likely to engage in OCB targeting their organization and (b) less likely to commit CWBs targeting their organization.

Hypothesis 1: Perceived union-targeted CSR will be (a) positively related to OCB-O, (b) negatively related to CWB-O, and (c) not significantly related to OCB-I or CWB-I.

Society-targeted CSR includes many items from the "discretionary" domain in Maignan et al.'s original scale and looks at topics such as charitable donations, philanthropic activities, and environmental sustainability. According to Rupp (2007), when workers perceive this to be high in their organizations, citizenship behaviors should increase while counterproductive work behaviors should decrease. This should operate as a perceived justice when workers "look out" into the world from their position in their organization. In response, helping behaviors should go up while undesirable behaviors go down. As was demonstrated by Colquitt (2001) among others, when workers perceive procedures to be fair, they are likely to respond with organizationally-directed outcome behaviors. Similar to Hypothesis 1, society-targeted CSR is a macro-level sub-dimension of stakeholder CSR, so is expected to be positively related to both sub-dimensions of OCB and negatively related to both sub-dimensions of CWB—but only significantly related to OCB-O and CWB-O.

Hypothesis 2: Perceived society-targeted CSR will be (a) positively related to OCB-O, (b) negatively related to CWB-O, and (c) not significantly related to OCB-I or CWB-I.

Counterproductive work behaviors are commonly exhibited when workers feel they have experienced injustice in the workplace. When workers feel as though their organization does not consider them an important stakeholder and treat them accordingly, workers are likely to respond with organizationally-directed behaviors. Consistent with Rupp (2007), different from Hypotheses 1 and 2, we expect perceptions of worker-targeted CSR to be both internalized (looking in) and externalized to peers (looking around); therefore, worker targeted CSR is expected to be positively related to both sub-dimensions of OCB and CWB.

Hypothesis 3: Perceived worker-targeted CSR will be (a) positively related to OCB-O and OCB-I and (b) negatively related to CWB-O and CWB-I.

Finally, when organizations are fiscally responsible, we expect that all stakeholders will respond favorably. At the worker level, this means that workers should be more likely to engage in helping OCB-Os that benefit the organization and less likely to commit CWB-Os that harm the organization. Similar to Hypotheses 1 and 2, owners and shareholders are likely to be perceived by individual workers as a macro/organizational-level group, so we do not expect to find a significant relationship between shareholder-targeted CSR and either OCB-I or CWB-I.

Hypothesis 4: Perceived shareholder-targeted CSR will be (a) positively related to OCB-O, (b) negatively related to CWB-O, and (c) not significantly related to OCB-I or CWB-I.

Methods—Study 1

The items used in Study 1 are adopted from the scale developed by Maignan et al. (1999). Note that the items have been adapted to be responded to by workers (the original scale was intended to be completed by managers). For example, one item from the original scale was worded: "My employer … is recognized as a trustworthy company," and in the adapted scale was reworded: "My employer is recognized by the public as a trustworthy company" and "My employer is recognized by workers as a trustworthy company." This is important because, in Study 2, these and additional items are further adapted to be more internally consistent. For the new stakeholder measure of CSR in Study 1, additional items are included that are adapted from other scales, including job attitudes, grievance negotiations, employer relationship.

Sample

Participants were 326 workers attending a week-long labor education program at a large Midwestern university. This sample included 72% males and 28% females, of which 25% were African American, 8% Latin American, 60% Caucasian, 1% Pacific Islander; the remainder reported other/bi-racial. The average age of participants was 44 years.

Procedure

The new scale includes the additional items adopted from measures of job attitudes, grievance negotiations, and relationship with the employer. Participants were given paper copies of questionnaires over a 4-day period and asked to complete during breaks in classroom activities. The participants were asked to complete surveys on a voluntary basis, with no other inducements. The items on the surveys were mostly directed toward either employers or unions.

Outcome Measures

All outcome measures were assessed with a 5-point Likert-type scale, with anchors ranging from 1 = *strongly disagree* and 5 = *strongly agree.*

Organizational Citizenship Behavior was measured with five items taken from Organ (1988) such as "I assist others with their work (when not asked)" and "I adhere to informal rules devised to maintain order." In study 1, only OCB-Os were measured, so OCB-I hypotheses were unable to be tested in this study.

Counterproductive Work Behavior was measured using a two-dimension scale from Bennett and Robinson (2000) comprised of eight total items: *Organizational* items included "Over the last 6 months how frequently have you intentionally worked slower?" and *Interpersonal* items included "Over the last 6 months how frequently have you bad mouthed the supervisor to others?" Given the low base rate of reported counterproductive work behaviors, items included (such as slacking at work or showing up late) were less extreme than those used in other studies (such as theft and sabotage behaviors).

Analysis—Study 1

We first conducted a confirmatory factor analysis using the structural equation modeling package *lavaan* in R. Using maximum likelihood estimation with the intercorrelation matrix for all of the items as input, we compared several a priori corporate social responsibility factor models, including a one-factor (general) model, a two-factor model suggested by exploratory factor analysis (EFA), and a four-factor model based on stakeholder CSR theory. Upon determining the best-fitting model (the four-factor model), using *lavaan*, we put the confirmatory model together with a structural path model complete with all factor loadings and path coefficients, demonstrating the hypothesized relationships between the Stakeholder CSR dimensions and the outcome variables. Table 1 includes the chi-square difference tests as well as fit indices, which indicate that the four-factor model is the best fit to the data of the models tested. The results of the chi-square difference tests show evidence that, given the data, the four-factor model is the best-fitting model with a p-value < .001.

TABLE 1

Study 1: χ^2 Difference Tests and Fit Indices for CSR Factor Structure

Models	df	AIC	BIC	χ^2	$\Delta\chi^2$	NFI	CFI	RMSEA
1-Factor (Null) Model	275	21643	21832	1181.5				
2-Factor Model	274	21522	21715	1058.3	123.18***	0.582	0.648	0.094
4-Factor Model	269	21235	21447	761.48	296.84***	0.699	0.779	0.075

***p < 0.001

Results—Study 1

Confirmatory Factor Analysis

We compared the fit of four different factor structures for Study 1, the first of which was a single-factor structure comprised of all 25 items from Maignan et al. The two-factor model then shows two sub-dimensions that are supported by theory; the first factor comprises all items related to either workers or their union, the second factor comprises all additional items, particularly regarding social obligations and shareholder responsibility.

Fit indices, found in Table 1, using exploratory factor analysis for the first study suggest that the best-fitting model was the four-factor structure, with items loading best onto the four sub-dimensions of stakeholder CSR: Shareholders,

Workers, Union, and Society. RMSEA for the four-factor model is .075, which is slightly higher than what many researchers consider to be a "good" fit [Kenny (2015) states that an RMSEA of .05 or lower implies a good-fitting model]. The CFI for the four-factor structure is .699, and the NFI is .779. While both below the threshold of .90 (the typical level implied by Bollen (1989b) to be a "good" fit), like the RMSEA both were the best fit to the data of all models tested. This was partially remedied in the second study, but future research in this area should pay much more attention to the development/refinement of this scale, including a more explicit measure of content validity, item reduction, and convergent/discriminant validation (see Shafiq et al. 2014 for more on scale development and stakeholder CSR). Lastly, the change in χ^2 from the two-factor model to the four-factor model was 296.84, and statistically significant at the $p<.001$ level, indicating that the change from the two-factor to the four-factor model was significant.

Counterproductive work behaviors, as mentioned, can be directed at either the organization or at individuals within the organization. Evidence for discriminant validity between these two dimensions of counterproductive work behaviors has been established in the literature on industrial and organizational psychology and is beyond the scope of this paper (e.g., Bennett and Robinson 1997). However, because these two dimensions are thought to be sub-facets of one overall construct of counterproductive work behaviors, the two are considered (and should be expected to be) highly correlated. Additionally, because counterproductive work behaviors and organizational citizenship behaviors have much in common, these constructs should all be expected to be correlated but should not share much variance. The results of the structural equation modeling for Study 1 show that these three constructs do indeed have a small amount of covariance, but the covariances are also statistically insignificant. The results of this can be seen in Tables 4 and 8 of the Appendix.

Hypothesis 2 predicted the relationships between society-targeted CSR (a) OCB-Os and (b) CWB-Os. The results of Study 1 do not provide support for Hypothesis 2. Hypothesis 3 predicted a positive relationship between worker-targeted CSR and OCB as well as a negative relationship between both sub-dimensions of CWB. There is evidence that shows no support for Hypothesis 3; in fact, the sign on the OCB dependent variable was significantly negative, the opposite of the hypothesized relationship. Shareholder CSR was predicted to be positively related to OCB-Os and negatively related to CWB-Os, and preliminary evidence also shows no support for this Hypothesis 4.

TABLE 2
Study 1: Full Model Regression Coefficients and (Standard Errors)

	CWB-O					CWB-I				
CSR–Shareholder	.032 (.067)				.181 (.137)	.115* (.065)				.221* (.132)
CSR–Society		–.038 (.066)			–.234* (.125)		.073 (.064)			–.106 (.117)
CSR–Union			.082 (.213)		.670** (.341)			.092 (.062)		.692** (.322)
CSR–Worker				.099 (.066)	–.546 (.116)				.110* (.062)	–.641* (.332)

	OCB				
CSR–Stockholder	.98 (.073)				.100 (.099)
CSR–Society		.074 (.066)			.162 (.152)
CSR–Union			.104 (.065)		.250** (.114)
CSR–Worker				.050 (.066)	–.306* (.180)

* p < .10; ** p < .05; *** p < .01

Study 2

Establishing model fit using factor analysis as well as predictive power is necessary for validating a measure (i.e., Chronbach and Meehl 1995). Using both exploratory and confirmatory factor analysis in Study 1, a stakeholder measure of CSR was shown not only to fit the data better than the original domain measure but also that the four-factor structure was the best fit to the data of the three models compared. Study 1 also shows evidence that the stakeholder measure had good predictive power for organizational outcomes. Study 2 served then as a replication and confirmation of the findings of Study 1 by keeping as many things about the study similar as possible, while modifying the measurement device to improve internal consistency. Recreating the CSR items from the Maignan et al. (1999) scale helped to again ensure a level of content validity, which is important when developing a new scale. As with Study 1, the same three outcomes were measured, and in Study 2 some additional items were included to increase internal consistency of the scales, to add an interpersonal/organizational dimension to the organizational citizenship behaviors construct.

Methods—Study 2

Sample

Participants were 400 workers attending a week-long labor education program at a large Midwestern university as was described in Study 1, albeit at a different point in time. This sample included 71% males and 25% females (the remainder did not report gender), of which 22% were African American, 8% Latin American, 1% Asian, and 62% Caucasian 1%, 1% Pacific Islander; the remainder reported other/bi-racial. The average age of participants was 42.6 years. Of the 400 surveys collected, 25 were dropped due to large levels of incompletion (75% or more incomplete), and three more were dropped due to inappropriate responses (profanities written in open-response items) or obvious carelessness in responding (all items answered "1"), so the final sample size for Study 2 was 372.

Procedure

Most items were only slightly modified versions of questions from the original questionnaires used in Study 1 in order to be more internally consistent, and are grouped similarly. Participants were given paper copies of questionnaires during their orientation to the labor education program and asked to complete. The participants were primed minimally, and instead simply asked to complete surveys on a voluntary basis. Just as in Study 1, the items on the survey were mostly directed toward either employers or unions.

Outcome Measures

All outcome measures were assessed with a 5-point Likert-type scale, with anchors ranging from 1 = *strongly disagree* and 5 = *strongly agree*. Respondents were provided with a 6th option, "I don't know," which was later coded as a neutral response ("Neither agree nor disagree").

Organizational Citizenship Behavior was measured with five items from Organ (1988) such as "I assist others with their work (when not asked)" and "I adhere to informal rules devised to maintain order." In addition to items from Study 1, additional items were included from Organ (1992)'s study and replications thereof. Additional items were from scales that looked at organizational citizenship behaviors as both interpersonally targeted and organizationally targeted (similar to the scale of counterproductive work behaviors used in Study 1, as well as here in Study 2).

Counterproductive Work Behavior was measured using a two-dimension scale from Bennett and Robinson (2000) comprised of eight total items: *Organizational* items included "Over the last 6 months how frequently have you intentionally worked slower?" and *Interpersonal* items included "Over the last 6 months how frequently have you bad mouthed the supervisor to others?"

Analysis—Study 2

Using the structural equation modeling package in R (*lavaan*, the same as in Study 1), we first began by running fit statistics on the newly developed scale, using maximum likelihood estimation with the intercorrelation matrix for all of the items as input. Having determined the best-fitting model in Study 1, we began the analysis with the new scale and put the confirmatory model together with a structural path model complete with all factor loadings and path coefficients, again demonstrating the hypothesized relationships between the Stakeholder CSR dimensions and the outcome variables.

Results—Study 2

Due to the high degree of collinearity between the Union CSR latent variable and the Worker CSR latent variable (.890), in Study 2, we were forced to drop one in order to run the analyses. We chose to drop the Union CSR variable; however, the Worker CSR variable if dropped produced nearly identical results (available upon request). The relationship between union-targeted CSR and CWB-O was hypothesized to be negative and the relationship with OCB-O to be positive. The relationship between worker-targeted CSR and (OCB-O and OCB-I were hypothesized to be positive and CWB-O and CWB-I to be negative. Results of Study 2 were all opposite from what was hypothesized, and all but one was statistically significant, thus unsupportive of Hypothesis 3. Hypothesis 2 was supported, and Hypothesis 4 was not. The full model with all regression coefficients, standard errors, and significance levels can be found in Table 3.

TABLE 3
Study 2: Full Model Regression Coefficients and (Standard Errors)

	OCB-O				OCB-I			
CSR–Stockholder	.114 (.073)			.058 (.110)	.082 (.066)			.060 (.100)
CSR–Society		.145** (.072)		.183* (.109)		.115* (.066)		.190* (.100)
CSR–Worker			–.026 (.069)	–.139* (.208)			–.097 (.063)	–.208** (.078)

* p < .10; ** p < .05; *** p < .01

	CWB-O				CWB-I			
CSR–Stockholder	.032 (.067)			.083 (.105)	.115* (.065)			.108 (.057)
CSR–Society		–.038 (.066)		–.175* (.105)		.073 (.064)		–.043 (.097)
CSR–Worker			.099 (.066)	.137* (.081)			.110* (.062)	.068 (.075)

* p < .10; ** p < .05; *** p <.01

A trade-off argument seems to suggest that, when workers feel their organization has violated social norms, they might respond by either *diminishing* helpful behaviors such as OCBs or *increasing* harmful behaviors such as CWBs and that engaging in one will offset the desire to engage in the other. Additionally, by targeting individuals within the organization, this behavior might offset the desire or need to target the organization with responsive behaviors.[3] Hypothesis 2 predicted the relationships between society-targeted CSR and (a) organizational citizenship behaviors and (b) counterproductive work behaviors. The relationship between society-targeted CSR and OCB directed toward the organization was positive and significant (β =.183*), and the estimated relationship between society-targeted CSR and CWB targeting the organization was negative (β = -.175*), and of the remaining two dependent variables, both coefficients were in the direction hypothesized, but one was statistically significant, thus supporting Hypothesis 2.

Hypothesis 3 predicted a positive relationship between worker-targeted CSR and both OCB-O and OCB-I, as well as a negative relationship between worker-targeted CSR and both CWB-O and CWB-I. This is due to the fact that, in accordance with Rupp (2011) perceptions of worker CSR can be thought of as both looking in ("I am a worker") and looking around ("we are all workers") might operate at multiple levels, and have multi-level outcomes. Results indicate partial support for the hypothesis. Both OCB-I and OCB-O were negatively related to worker CSR (β = -.208**, β = -.139*, respectively), and both CWB-I and CWB-O were positively related to worker CSR (β = .068, β = .137*). One explanation for these results is that perhaps to some extent causality flows in the opposite direction of that hypothesized and that, in reality, low levels of OCB might cause employers to be more socially responsible toward their workers, and given that these data are cross-sectional, definitively arguing for causality is challenging. Taken together, it could also be the case that as CSR targeting workers increases, workers are less likely to commit CWBs, and this is offset by a decrease in citizenship behaviors as well. Considering the paradox discussed above, a trade-off between citizenship behaviors and other types of performance, such as CWB, might exist—and could be the driver behind the results seen here.

Finally, Hypothesis 4 predicted the relationship between shareholder CSR and organizational citizenship behaviors directed toward the organization to be positive. The results of Study 2 showed no support for Hypothesis 4. The slope coefficient for shareholder CSR on CWB-O (β = .083) and OCB-O (β = .058), but neither were significant. Additionally, the results of Study 2 indicate that the relationship between shareholder CSR and CWB-I was positive and significant (β = .108*, $p<.10$) and shareholder CSR and OCB-I was positive but not significant (β = .060).

Discussion, Future Directions, and Conclusion

The results of the data analyses in Study 1 show early indications that a stakeholder model and measure of CSR are a somewhat effective way to understand and operationalize the construct of perceived stakeholder CSR in a way that not only fits the data well but has predictive capabilities as well. As the stakeholder perspective on CSR continues to gain ground in organizational studies, an increasing need for a validated measure of CSR through the lens of stakeholders grows increasingly necessary. As can be seen from the results of Study 1, confirmatory factor analysis indicates that the stakeholder measure of CSR offered in this paper is a good fit to the data, moreover than the domain CSR measure used in previous research.

In the cases of both worker-targeted CSR (in both studies) and union-targeted CSR (in Study 1), and the outcomes counterproductive work behaviors and organizational citizenship behaviors, the mixed results cast a wider shadow on the relationship between these four constructs and present one limitation of this study. Given that both studies are cross-sectional in nature, there is no definitive way to establish causality. As such, it is possible that worker and union-targeted CSR do not *cause* counterproductive work behaviors and citizenship behavior, but perhaps that *causality flows in the opposite direction*, or perhaps that causality flows bi-directionally. One interpretation of this would be that in some instances, high levels of counterproductive work behaviors and lower levels of organizational citizenship behaviors cause organizations to engage in higher levels of CSR directed toward workers, unions, and society, while in other instances when organizations engage in higher levels of CSR targeting those stakeholder groups, organizational citizenship behaviors are higher, and counterproductive work behaviors are lower as a result. Some have argued that OCB operates proactively, and CWB reactively to various environmental stimuli and/or individual differences, which may also explain why workers' reactions to their organization's various targeted CSR initiatives and the mixed results of these studies.

Study 1 shows preliminary evidence with respect to society-targeted CSR and CWB-Os and OCB-Os. Study 2 investigated these findings in more detail, and the results demonstrate that society-targeted CSR as a sub-dimension of the stakeholder model of CSR appears to have good predictive power in determining a variety of organizational outcomes. These findings support the view that when workers view their organization as socially responsive to others outside of the company, they respond with overall favorable behaviors. These findings also support the argument that OCB is a proactive, rather than a reactive, construct and that organizations respond favorably to helpful citizenship behaviors with socially responsible behaviors. In other words, when workers are "looking out" into the external environment and perceiving their organization as doing the same and exhibiting positive socially responsible behaviors, their extra-role behaviors mirror those. With the collection of longitudinal data, it is possible to unpack this more thoroughly and to uncover evidence that of the four CSR stakeholders being examined in this study, society might have the strongest predictive power. Future research should examine this further, and perhaps consider breaking down the society stakeholder group further, into sub-groups such as non-governmental organizations (NGOs), the ecological environment, and other third-party stakeholders not directly involved in the organizational operations yet are affected by them all the same.

The findings in this study are the first of its kind to demonstrate empirically how the stakeholder dimensions of CSR can predict behavioral outcomes. As the results indicate, the hypotheses were somewhat supported, but the ways in which they were not supported is perhaps just as interesting as the ones that were. The results of this research are broad and hold some serious implications for organizational practices and management. First, as scholars and managers alike have known for some time, CSR perceptions have strong predictive power in determining several types of worker behaviors. Additionally, as Rupp and colleagues (2008) have suggested, organizations must not only consider workers attitudes toward internalized justice perceptions, but organizations must also look outside of their walls in considering implications for organizational success. These studies show evidence that workers are not only concerned with justice and fairness directed at themselves but at others as well. Of equal importance, these studies show evidence that, indeed, different stakeholders *do* matter differently to workers. This means that organizations would do well to consider how their workers perceive their fair treatment of other stakeholders besides just workers and shareholders, and when workers perceive their organization as fair, this climate can act as a determinate of both desirable and undesirable behaviors.

Research on corporate social responsibility continues to grow among management, practitioners, consultants, and scholars. Studying how CSR directed toward specific stakeholders operates, how it is supported by organizations and workers, and how it influences organizational outcomes should only continue to grow as more organizations embrace social justice initiatives, sustainability, and socially responsible behaviors. This paper begins to explore a previously under-examined relationship between CSR and organizational behavioral outcomes and provides several avenues for future research to explore. The future of CSR from the stakeholder perspective, as well as labor relations and worker attitudes toward CSR targeting specific stakeholder groups, appear to be quite bright.

Endnotes

[1] Dahlsrud (2008) provides an excellent summary of the literature on stakeholder CSR definitions.

[2] These authors discuss the pushback against the Canadian Labour Association of Canada for negotiating "sweetheart contracts" with several employers, resulting in weak CBAs representing union members.

[3] In a similar vein, Bolino, Klotz, and Turnley (2012) found that the trade-off of stress in personal/family life caused by engaging in OCBs with the professional benefits was worth it to workers.

References

Aguilera, R. V., Rupp, D. E., Williams, C. A., and Ganapathi, J. (2007"). Putting the S Back in Corporate Social Responsibility: A Multilevel Theory of Social Change in Organizations." *Academy of Management Review*, 32(3), 836-863.

Arbuckle, J. L., and Wothke, W. (1999). Amos 4.0. Chicago, IL: Smallwaters.

Balser, D., and Winkler, A. (2012). "Worker Behavior on the Job: A Multi-Methods Study of Labor Cooperation with Management." *Journal of Labor Research*, *33*(3), 388-413.

Bazillier, R., and Vauday, J. (2009). The Greenwashing Machine: Is CSR More Than Communication? DR LEO ; 2009-10. DR LEO, 2009-10. 2009. <hal-00448861v3>

Bemmels, B. (1994). "The Determinants of Grievance Initiation." *Industrial and Labor Relations Review* [serial online]. 47(2):285-301. Available from Business Source Complete, Ipswich, MA.

Robinson, S. L., and Bennett, R. J. (1997). Workplace Deviance: Its Definition, Its Manifestations, and Its Causes. In Lewicki, R. J., Bies, R. J., and Sheppard, B. H., eds., *Research on Negotiations in Organizations*, Vol. 5, pp. 3-27. Amsterdam: Elsevier Science.

Bennett, R. J., and Robinson, S. L. (2000). "Development of a Measure of Workplace Deviance." *Journal of Applied Psychology*, 85(3), 349.

Bentler, P. M. (1990). "Comparative Fit Indexes in Structural Models." *Psychological Bulletin*, *107*(2), 238.

Bergeron, D. M. (2007). "The Potential Paradox of Organizational Citizenship Behavior: Good Citizens at What Cost? "*Academy of Management Review*, *32*(4), 1078-1095.

Bergeron, D. M., Shipp, A. J., Rosen, B., and Furst, S. A. (2011). "Organizational Citizenship Behavior and Career Outcomes: The Cost of Being a Good Citizen." *Journal of Management, 39*(4).

Berman, S. L., Wicks, A. C., Kotha, S., and Jones, T. M. (1999). "Does Stakeholder Orientation Matter? The Relationship Between Stakeholder Management Models and Firm Financial Performance." *Academy of Management Journal*, *42*(5), 488-506.

Bies, R. J., and Moag, J. S. (1986). Interactional Justice: Communication Criteria of Fairness. Research on Negotiation in Organizations, Vol. 1, pp. 43-55. Amsterdam: Elsevier Science.

Bolino, M. C., Klotz, A. C., Turnley, W. H., and Harvey, J. (2013). "Exploring the Dark Side of Organizational Citizenship Behavior."*Journal of Organizational Behavior*, *34* (4), 542-559.

Bollen, Kenneth A. (1989a). *Structural Equations with Latent Variables.* New York : Wiley.

Bollen, K. A. (1989b). "A New Incremental Fit Index for General Structural Equation Models." *Sociological Methods and Research*, *17*(3), 303-316.

Burnham, K. P., and Anderson, D. R. (2004). "Multimodel Inference Understanding AIC and BIC in Model Selection." *Sociological Methods and Research*, *33*(2), 261-304.

Camplin, J. C. (2011). "Aligning Safety andSocial Responsibility: Safety as a Value-Add to Corporate Stakeholders." *Professional Safety*, *56*(5), 46-55.

Carpenter, N. C., Berry, C. M., and Houston, L. (2014). "A Meta-Analytic Comparison of Self- Reported and Other-Reported Organizational Citizenship Behavior." *Journal of Organizational Behavior*, *35*(4), 547-574.

Carroll, A. B. (1979). "A Three-Dimensional Conceptual Model of Corporate Performance." *Academy of Management Review*, *4* (4), 497-505.

Che, X. (2012). An Exploratory Study of Reception of Organizational Citizenship Behaviors and Work Related Outcomes: It is Good for Your Co-workers. (Unpublished doctoral dissertation). University of South Florida. scholarcommons.usf.edu/etd/4297/

Colquitt, J. A. (2001). "On the Dimensionality of Organizational Justice: A Construct Validation of a Measure." *Journal of Applied Psychology*, 86(3), 386.

Dahlsrud, A. (2008). "How Corporate Social Responsibility Is Defined: An Analysis of 37 Definitions." *Corporate Social Responsibility and Environmental Management*, *15*(1), 1-13.

Dar, O. L. (2010). "Trust in Co-Workers and Employee Behaviours at Work." *International Review of Business Research Papers*, *6*(1), 194-204.

Davis-Kean, P. E. (2005). "The Influence of Parent Education and Family Income on Child Achievement: The Indirect Role of Parental Expectations and the Home Environment." *Journal of Family Psychology*, *19*(2), 294.

Dean, L. R. (1954). "Union Activity and Dual Loyalty." *Industrial and Labor Relations Review*, 526-536.

DeVellis, R. F. (2012). *Scale Development: Theory and Applications*, Vol. 26. Thousand Oaks, CA: Sage.

Ferris, D. L., Brown, D. J., Berry, J. W., and Huiwen, L. (2008). "The Development and Validation of the Workplace Ostracism Scale." *Journal of Applied Psychology*, *93*(6), 1348-1366.

Fleishman, E. A., and Harris, E. F. (1962). "Patterns of Leadership Behavior Related to Employee Grievances and Turnover." *Personnel Psychology*, *15*(1), 43-56.

Fodchuk, K. M. (2007). "I. Management Principles: The Theory of Management: Work Environments That Negate Counterproductive Behaviors and Foster Organizational Citizenship: Research-Based Recommendations for Managers." *Psychologist–Manager Journal*, *10*(1), 27-46.

Fox, S., Spector, P. E., and Miles, D. (2001). "Counterproductive Work Behavior (CWB) in Response to Job Stressors and Organizational Justice: Some Mediator and Moderator Tests for Autonomy and Emotions." *Journal of Vocational Behavior*, 59(3), 291-309.

Freeman, R. E. (1984). "Stakeholder Management: Framework and Philosophy." Mansfield, MA: Pitman.

Gond, J. P., El-Akremi, A., Igalens, J., and Swaen, V. (2010). *Corporate Social Responsibility Influence on Employees.* Research Paper Series. International Centre for Corporate Social Responsibility. Nottingham University.

Groves, K., and LaRocca, M. (2011). "Responsible Leadership Outcomes Via Stakeholder CSR Values: Testing a Values-Centered Model of Transformational Leadership." *Journal of Business Ethics* [serial online], *98*, 37-55.

Groves, K. S. (2014). Examining Leader–Follower Congruence of Social Responsibility Values in Transformational Leadership. *Journal of Leadership and Organizational Studies*, *21*(3), 227-243.

Hamann, R., and Kapelus, P. (2004). "Corporate Social Responsibility in Mining in Southern Africa: Fair Accountability or Just Greenwash?" *Development*, 47(3), 85-92.

Hammer, T. H., and Stern, R. N. (1986). "A Yo-Yo Model of Cooperation: Union Participation in Management at the Rath Packing Company." *Industrial and Labor Relations Review*, *39*(3), 337-349.

Harcourt, M., and Lam, H. (2007). "Union Certification: A Critical Analysis and Proposed Alternative." *WorkingUSA*, *10*(3), 327-345.

Hensley, R. L. (1999). "A Review of Operations Management Studies Using Scale Development Techniques." *Journal of Operations Management*, *17*(3), 343-358.

Hinkin, T. R. (1998). "A Brief Tutorial on the Development of Measures for Use in Survey Questionnaires." *Organizational Research Methods*, *1*(1), 104-121.

Hinkin, T. R. (2005). "Scale Development Principles and Practices." *Research in Organizations: Foundations and Methods of Inquiry*, 161-179.

Kaplan, H. B. (1975). *Self-Attitudes and Deviant Behavior.* Pacific Palisades, CA: Goodyear.

Kelly, J., and Kelly, C. (1991). "Them and Us: Social Psychology and The New Industrial Relations." *British Journal of Industrial Relations*, *29*(1), 25-48.

Kenny, D. A. (2012). *Measuring Model Fit.* davidakenny.net/cm/fit.htm

Klotz, A. C., and Bolino, M. C. (2013). "Citizenship and Counterproductive Work Behavior: A Moral Licensing View." *Academy of Management Review*, *38*(2), 292-306.

Labig, C. E., and Greer, C. R. (1988). "Grievance Initiation: A Literature Survey and Suggestions for Future Research." *Journal of Labor Research*, *9*(1), 1-27.

Leana, C. R., Ahlbrandt, R. S., and Murrell, A. J. (1992). "The Effects of Employee Involvement Programs on Unionized Workers' Attitudes, Perceptions, and Preferences in Decision Making." *Academy of Management Journal*, *35*(4), 861-873.

Maignan, I., Ferrell, O. C., and Hult, G. T. M. (1999). "Corporate Citizenship: Cultural Antecedents and Business Benefits." *Journal of the Academy of Marketing Science*, 27(4), 455-469.

McClendon, J. A., Wheeler, H. N., and Weikle, R. D. (1998). "The Individual Decision to Unionize." *Labor Studies Journal*, 23(3), 34-54.

McWilliams, A., Siegel, D. S., and Wright, P. M. (2006). "Corporate Social Responsibility: Strategic Implications." *Journal of Management Studies*, *43*(1), 1-18.

Organ, D. W. (1988). *Organizational Citizenship Behavior: The Good Soldier Syndrome.* Lexington, MA: Lexington Books.

Orlitzky, M., Schmidt, F. L., and Rynes, S. L. (2003). "Corporate Social and Financial Performance: A Meta-Analysis." *Organization Studies*, 24(3), 403-441.

Price, J. L., and Mueller, C. W. (1986). *Handbook of Organizational Measurement.* Marshfield, MA: Pitman.

Rolland, D., and O'Keefe Bazzoni, J. (2009). "Greening Corporate Identity: CSR Online Corporate Identity Reporting." *Corporate Communications: An International Journal*, *14*(3), 249-263.

Rupp, D. E. (2011). "An Employee-Centered Model of Organizational Justice and Social Responsibility." *Organizational Psychology Review*, 1(1), 72-94.

Rupp, D. E., Shao, R., Jones, K. S., and Liao, H. (2014). "The Utility of a Multifoci Approach to the Study of Organizational Justice: A Meta-Analytic Investigation into the Consideration of Normative Rules, Moral Accountability, Bandwidth-Fidelity, and Social Exchange." *Organizational Behavior and Human Decision Processes*, 123(2), 159-185.

Shafiq, A., Klassen, R. D., Johnson, P. F., and Awaysheh, A. (2014). "Socially Responsible Practices: An Exploratory Study on Scale Development Using Stakeholder Theory." *Decision Sciences*, *45*(4), 683-716.

Sweeney, P. D., and McFarlin, D. B. (1993). "Workers' Evaluations of the 'Ends' and the 'Means': An Examination of Four Models of Distributive and Procedural Justice." *Organizational Behavior and Human Decision Processes*,55(1), 23-40.

Tepper, B. J., and Taylor, E. C. (2003). "Relationships Among Supervisors' and Subordinates' Procedural Justice Perceptions and Organizational Citizenship Behaviors." *Academy of Management Journal*, 46(1), 97-105.

Werther, W. B., and Chandler, D. (2005). "Strategic Corporate Social Responsibility as Global Brand Insurance." *Business Horizons*, *48*(4), 317-324.

Werther Jr, W. B., and Chandler, D. (2010). *Strategic Corporate Social Responsibility: Stakeholders in a Global Environment.* Thousand Oaks, CA: Sage.

Williams, L. J., and Anderson, S. E. (1991). "Job Satisfaction and Organizational Commitment as Predictors of Organizational Citizenship and In-Role Behaviors." *Journal of Management*, 17(3), 601-617.

Woods, C. M. (2006). "Careless Responding to Reverse-Worded Items: Implications for Confirmatory Factor Analysis." *Journal of Psychopathology and Behavioral Assessment*, *28*(3), 186-191.

XII. LERA Annual Reports

LERA Executive Board Meeting Minutes

Tuesday, January 5, 2016, 11 a.m. PT
In Person, and Via Conference Call

Call to order—the meeting was called to order at 11 a.m. by Bonnie Castrey, President. Present at the meeting were officers President Bonnie Castrey and Secretary-Treasurer Craig Olson, NCAC Chair William Canak and President Elect Janice Bellace. Also present were board members John Budd (Program Committee Co-Chair), Alex Colvin, Dennis Dabney, Tazewell Hurst III, Ken May, and Patrice Mareschal. Unable to attend were Janice Bellace, Dennis Dabney, Rebecca Givan, Charles Jeszeck, Kris Rondeau, and Beth Schindler. LERA staff in attendance was Emily Smith. Via Conference Call – Saul Rubinstein, Marlene Heyser, Bill Dwyer, Owen Herrnstadt, Rebecca Givan, Jeff Keefe.

Reports

President and Executive Committee Report—President Bonnie Castrey reports the LERA 68th Annual Meeting is going well. Appointments to LERA Awards Committee (Tia Denenberg and Doug Kruse), the Development Committee (Jim Pruitt and Cyndi Furseth), have been made, and she continues to review other committees. Officers have remained in close contact with LERA staff throughout the last six months. Many of our software and files were deleted and they have been salvaged or reconstructed.

Finance Report—Secretary-Treasurer Craig Olson reports that we need an additional layer of oversight on cash and deposit handling so that there is no duplication in cash-handling. Motion to accept audit moved by Alex Colvin and seconded by David Lewin, and unanimously passed, no no's and no abstentions. Olson reports that 2015 was not a good year for the association, at a loss of $65000, not the worst deficit in recent years, but unsustainable. We had a drop in dues income, about 15%. Our sponsorship income is down 33%. There are user issues with our online MemberClicks software complicating online renewals. Our member losses have stemmed with late Fall membership promos, but we are still down 8% overall. We are down 15% in membership revenue overall. Overall, looking at membership revenues, we lost about $20,000 in income between organizational members, member dues, and sponsorship income. The other factor for loss is the implementation and planning of the Pittsburgh Annual Meeting. The former director over-ordered meals and hospitality events, accounting for a $35,000 swing from 2014 to 2015 in meetings. Our publications were also more expensive in 2015, but we have reduced our professional staff to one so 2016 is expected to see a small surplus. We will augment with part-time and student help in 2016. Alex Colvin commends President Bonnie Castrey and Secretary-Treasurer on their handling of the budget crisis that occurred this year, bringing the organization back into fiscal responsibility. Bill Canak moves to accept financials, seconded by Alex Colvin. The motion passed unanimous, no no's or abstentions.

Strategic Thinking Committee—David Lewin reports that a Request for Proposal has been formulated to review the headquarters home for the organization. Does it continue to make sense for LERA to continue its headquarters at the University of Illinois at Urbana-Champaign? A letter of interest was formulated to see if there was any interest in other Universities becoming LERA's new home, and several Universities expressed interest. (Rutgers, Cornell, Univ. of Illinois, and others have responded). We ought to consider continuing at the University of Illinois for the time being, a year or two, and then consider a new home in a few years. Our current thinking is that we ought to continue at the University of Illinois at Urbana-Champaign. There have been leadership changes and faculty turn-over at the University of Illinois – five new faculty this fall, two more faculty next fall, and just interviewed for a third position. We have had some excellent professors move, but have replaced them with young and exciting new assistant professors. When we began the review, we were examining what our best practices would be with LERA's headquarters and organization. Rutgers, Cornell, Penn State, and University of Illinois at Urbana-Champaign have all expressed interest. It's important that we stabilize the organization over the next few years. We would like to encourage the new professors at the University of Illinois School of LER become LERA members and that a local chapter be formulated in Central Illinois (William Canak). Janice Bellace comments that other schools are happy to host LERA, but they did not wish to take it away from Illinois if University of Illinois was interested in retaining LERA. Rebecca Givan reports the same for Rutgers; they were willing and able to step up, but not wishing to compete with Illinois. Perhaps we could make it a requirement that the institution needs to have a local chapter. We could use a few year time frame to require a local chapter to be part and parcel with the headquarters institution. Saul Rubinstein suggests that multiple Universities share in the support of LERA (consortium in this digital age) – not necessary that we could have certain capacities supported by more than one organization. There are discreet things that other Universities could contribute you (ILR Press, etc.) marketing. Owen Herrnstadt agrees that we need one administrative home with a good amount of support, but it is a great idea to develop the idea of getting more direct involvement by the other labor centers and schools around the country. Hopefully, some of them are represented by the board, but an increased institutional presence would be great. There are aspects of what LERA does or could be doing that could be linked and enhanced with digital cooperative effort. Bill Canak suggests that we continue to consider the aspect of multiple universities contributing, what would be the complexities and what would be the efficiencies. We could put a group together to think on this. Motion to continue to study this topic, for the next two years, and re-consider sending an RFP in two years at the end of 2017. Motion by Ken May, seconded by Janice Bellace, passed unanimously, no no's and no abstentions.

National Chapter Advisory Committee—William Canak reports that we had a successful showing of chapter awards in 2015, which is great. He also reports that there are activities happening at local chapters that may have repercussions at LERA, including CEU's and SHRM credits that happen routinely at chapters, why not LERA? We also have video happening at local chapters, as well as shirts, mugs, and other LERA branded items. Everyone who registers at TERRA conference, everyone gets a 4 G thumb drive that contains all papers and related documents that people would like to share for LERA meeting attendees. Chapters do this, why not LERA. We have been asking Chapters to promote LERA more, joint membership, and polling constituencies about why various organizations join or do not join. These issues are discussions (chapter administration, chapter programming, chapter marketing and participation) that happen at the LERA Annual Meeting at the chapter representatives meeting. What are best practices on administration, membership, and programming? Hoping to reproduce our awards success in 2016. SW Pennsylvania, Florida, new Student Chapter – currently pushing Universities to pursue student LERA chapters, an official chartered student chapter. In Minneapolis, we will be doing a workshop on how to create a student chapter. We might consider finding out which Universities that do not have an ILR school might be targeted (Janice Bellace). Bill and Janice to follow-up as new membership business.

Development Committee—Marlene Heyser Committee chair reports. Bonnie has appointed two new members, Jim Pruitt and Cyndi Furseth, and we will put together some regularly scheduled conference calls for the Development Committee. We need a new strategy for sponsorship dollars. We need an on-going revenue stream. Bonnie Castrey thanks Marlene for her continued efforts. Janice Bellace, President sends out a letter in the Fall. Would there be a way to incorporate the already established Industry Councils and Interest Sections to reach out on an institutional level to the organizations that make the most sense? We will need to revitalize the Industry Councils and Interest Sections in order to do this. We need to address membership in a very real way where we have a meeting specifically to address membership needs – David Lewin made this suggestion. Steve Sleigh and a few others like him should be at the table when we put together this meeting.

Membership Report—Jeffrey Keefe reports that his efforts to support academic enrolment by adding a LERA Journal has hit a wall. The editorial committee is, however, pursing a joint publication with the ILR Review.

LERA 68th Annual Meeting Program Committee—Bonnie Castrey reports that work continues apace. A P&L has been put together for the meeting, and the program is nearly in place. We have worked with the ISA to do a joint program on the Thursday, May 26, 2016 of the program. Registration prices accepted by the board with the exception to increase on-site registration for members to $375. The student rate should also be listed as the student/apprentice rate. The motion passed unanimously, no no's and no abstentions. Bonnie Castrey reports that NAA will conduct a joint venture, a LERA/NAA session on the LERA program and a LERA/NAA session on the NAA program at their meeting. Allen Ponak requests that we would like to waive registration fees for the respective registrants at the respective meetings.

LERA 69th Annual Meeting—Janice Bellace reports that Alex Colvin has spearheaded a joint meeting in 2017 with CIRA the Canadian Industrial Relations Association. This might mean additional LERA members.

LERA Winter 2016 Meeting—Sanford Jacoby submitted a report about the activities of the program committee leading up to the LERA Winter 2016 Meeting in San Francisco, CA January 3-5, 2016.

LERA will establish a LERA ILR Review moved and seconded, passes unanimously. LERA would like to establish a book review. The Board thinks that it would be difficult to administrate this award and motions that we pursue that we recognize leading book authors at the LERA Annual Meeting, motioned by Ken May and seconded by Patrice Mareschal.

New Business

LERA Bylaw Revisions—These are still being reviewed by the board and we will revisit at the February board meeting, 2016.

LERA Annual Meeting swag, shirts, and videos are a good idea. We will try to pursue these at the next LERA Annual Meeting. Would it be good to conduct brief interviews with meeting attendees about why they attend and what issues are important to them. Bill Canak and John Budd to work on this jointly, possibly with some student help.

Next meeting will take place via conference call in February, closed executive session. We also need to conduct a session on membership. Steve Sleigh and Jeff Keefe should be there.

Adjournment—The conference call meeting was adjourned at 2:10 p.m. by President Bonnie Castrey. Ken May moved the motion to the floor and Patrice Mareschal seconded it. The motion was unanimously passed by the board.

LERA Executive Board Meeting Minutes

Friday, February 19, 2016, 10 a.m. EST
Via Conference Call

Call to order—the meeting was called to order at 10 a.m. by Bonnie Castrey, President. Present at the meeting were officers President Bonnie Castrey, Secretary-Treasurer Craig Olson, and NCAC Chair Bill Canak. Also present were board members John Budd (Program Committee Co-Chair), Eric Fidoten, Owen Herrnstadt, Tazewell Hurst, III, Ken May, Kris Rondeau, Janice Bellace (President Elect), Marty Mulloy (Past President). Unable to attend were Janice Bellace, Dennis Dabney, Rebecca Givan, Charles Jeszeck, Kris Rondeau, and Beth Schindler. LERA staff in attendance was Emily Smith.

Reports

President and Executive Committee Report—President Bonnie Castrey reports that past director Eric Duchinsky tendered his resignation effective August 15, 2015. The Executive Committee met, reviewed the budget and financials, and appointed Emily Smith as Interim Executive Director, with a subsequent increase in pay. This motion came forward to the board as a committee action, was discussed, and ratified unanimously by the Board. The process of reviewing the budget determined that profit and loss are not meeting projections for 2015, and the officers of the organization continue to review the situation actively.

Interim Director Report—Emily Smith reports that all administrative functions are complete for 2015, such as audit, insurance, and filings. The Research Volume has been mailed to members and Perspectives on Work proceeds according to schedule to be mailed this Fall. Top priority for remainder of 2015 is to increase member dues revenue and bring back the members we have lost if possible. We are planning five member promotions, three in September and two in October to help bring up member dues for 2015. Essentially, revenue is down in all categories, including new members, renewals, library subscribers, organizational members, and meeting income. Much of this may be linked to a new fulfillment system put in place in 2014, but additional information is needed and a full discovery will be made. Most areas of expense are up in 2015 compared to the projected budget, including Annual Meeting, *Research Volume*, *Perspectives on Work*. Steps will be taken in 2016 to bring expenses back within projections. Our short-term remedy for the losses sustained in 2015 is maintain a full-time staff of one at the LERA office instead of two, and augment that with temporary clerical help as needed, for at least the period of a year. Long-term decisions will need to be made after an inquiry is complete about the future of LERA's headquarters.

Finance Report—Secretary-Treasurer Craig Olson reports that member dues are down precipitously in 2015. This may be due to the change from fulfillment through the UIP to MemberClicks. Members, library subscribers, and contributions are all down, likely connected to issues with the MC software or the user experience of it. The Annual Meeting income was also down from last year, because the meeting had 100 fewer attendees from 2014 to 2015. Almost all areas of expense are up, and need to be considered and controlled in 2016. We will keep full-time staff members to one at LERA for the next year to help with this.

Annual Meeting Profit and Loss—Secretary-Treasurer Craig Olson reports that the organization lost money on the LERA 67th Annual Meeting, May 28-31, 2015, in Pittsburgh, PA. This was carefully reviewed by staff and the causes were over-ordering of food and beverage, ordering additional breaks (food and beverage events other than meals or receptions), paying for travel and lodging of some meeting participants, and paying for registration of some meeting participants. The last three years of profit and loss for LERA Annual Meetings have been reviewed, and steps will be taken in 2016 to ensure that we have a profitable meeting. The contracts for Minneapolis and Anaheim will be reviewed by staff and altered where needed and if possible. The Minnesota venue appears to be affordable, while the Anaheim venue may not be according to the contracts as they currently stand. Some discussion took place about the attributes of setting separate prices for registration and various events, versus having a registration price that covers all events (or most of them) at the meeting. Having a reduced registration price can be attractive to some. Having separate tickets for hospitality events can reduce inclusivity at the conference and increase staffing. This will continue to be evaluated.

LERA 68th Annual Meeting, May 26-29, 2016, Minneapolis Update—Bonnie Castrey reports that Emily Smith and she will be completing a site visit at the Millennium Hotel by the end of September, and reviewing the contract that we have in place. We have printed a postcard to promote the meeting and John Budd has done considerable advance promotion already, including at the Association of Labor Relations Agencies (ALRA) conference, which just took place this summer in Minneapolis. Marketing promotion begins in earnest when the program is set in January 2016.

In Minneapolis, LERA will meet jointly with the Industry Studies Association, with one day overlapping between the two, as we did in Portland 2014. Both organizations are working towards making this a collaborative day. President Bonnie Castrey and Emily Smith will be meet with the ISA program chairs to discuss the joint meeting in Minneapolis. Thursday, May 26 will be an overlap day and we will be discussing programming, joint plenaries, other collaborative events, hotel logistics, registration, and more.

Program Committee Co-Chair John Budd reports that Thursday and Friday will be the key days to pull in local practitioners, something to consider as Thursday is already constrained as the overlap day between ISA and LERA. He also reports that this conference is scheduled for the Memorial Day Weekend, and that could prove challenging to bringing in locals who are making vacation plans.

Executive Session

New LERA Executive Director—The LERA Board excused LERA staff and conducted a closed executive session of the board meeting. President Bonnie Castrey reported that Emily Smith has been appointed as LERA Executive Director with a salary of $75,000 per year.

Next Board Meeting—We are currently dealing with a number of challenges, and President Bonnie Castrey encourages all board members to attend the next board meeting in person for follow-up to these discussions, potential bylaw changes, and important issues facing the organization including strategic thinking committee findings. It will take place Tuesday, January 5, 2016, at the PARC 55 San Francisco, Market Street Room, and begin at 11am. A call-in option will be explored for those who cannot make the meeting in person; **however, this will be a crucial meeting for this Board and all are encouraged to attend in person**.

Adjournment—The conference call meeting was adjourned at 10:22 a.m. by President Bonnie Castrey.

LERA Executive Board Meeting Minutes

Friday, April 15, 2016, 10 a.m. to 4 p.m. CT

Big Ten Conference Center, 5440 Park Place, Rosemont, IL 60018

Call to order—the meeting was called to order at 10 a.m. by Bonnie Castrey, President. Present at the meeting were officers President Bonnie Castrey, President Elect Janice Bellace, Past President Marty Mulloy, and Secretary-Treasurer Craig Olson. Also present were board members John Budd (Program Committee Co-Chair), Charles Jeszeck, and Ken May. LERA Staff Emily Smith was present. Attending via conference call were Tazewell Hurst, III, Bill Dwyer, Bill Canak, Dennis Dabney, Eric Fidoten, Marlene Heyser, and Patrice Mareschal. Others in attendance were Jeff Keefe (Membership Committee Chair), David Lewin (Strategic Thinking Committee), and Stephen Sleigh. Unable to attend were Alex Colvin, Rebecca Givan, Owen Herrnstadt, Saul Rubinstein, Kris Rondeau, Beth Schindler. LERA staff in attendance was Emily Smith.

Reports

President and Executive Committee Report—President Bonnie Castrey reports the board decided to conduct a special board meeting centered around membership at the January 2016 board meeting, and the meeting taking place today is that meeting.

LERA 68th Annual Meeting 2016 Update—President Bonnie Castrey reports that registration is tracking well, in terms of both numbers of registrants and in revenue received.

LERA Finance & Membership Committee—Craig Olson, Secretary-Treasurer and Jeff Keefe, Chair, report on the state of membership in the LERA. This working meeting business and brain storming session revolved around discussion of increasing membership in the Association including current membership numbers and goals for 2016 and beyond.

New Business

Methods of reaching out to diverse constituencies were discussed so that our events, membership, and leadership reflect diverse demographics of the workforce. Two key methods were discussed to increase membership and ties to local chapters, as well as labor and management.

Creating Ties With Local Chapters—The first method is to address our bylaws to create Affiliate status for local LERA chapter members. This would require a bylaws change so that we can more closely tie the LERA chapters to LERA, and one of these changes would include the addition of Regional Vice Presidents that will stand for election to the LERA board, from three regions across the country to represent local LERA chapters. The three regions (East, West, and MidWest) would rotate candidates between labor, management, and neutral/government/academic, so that one of each perspective serve in any given year. For nominal dues from the chapter, the LERA will include local chapter members in our e-Bulletins and give them member rates at our events. They would not have additional member benefits.

Creating Ties with Labor and Management—Also, the LERA will augment organizational member benefits to non-academic organizational members by offering both Associate status to non-academic organizational members' employees, up to 10-12, and additional skill-building training of their choice that will lead to a certificate of completion. These individuals, as Associates, would be eligible to receive our e-Bulletins, and special member pricing at LERA events. In addition, the LERA can offer to put together special pre-conference (Annual Meeting) day training workshops for their employees, on topics that they consider valuable as part of their organizational membership package. Also, programming could take place with local chapters. We might also offer Industry Council advisory positions to organizational members.

A $500 organizational membership would yield: 1 Full Regular Individual Membership, Recognition (web, pubs, meeting signage), access to publications, print and online; 10-12 affiliates who will receive e-Bulletin, member pricing at events, and online access to publications.

Assignments were made for individual member of this meeting to follow up with specific individuals and organizations. A list of brainstorming follows:

- *Labor Economists*—While LERA's number of labor economists has decreased, the Society of Labor Economists has grown. We should ensure that we market the fact that we have 18 sessions at the AEA meetings to entice them, and be sure to mention that you must be a member to submit a proposal. Craig Olson suggests that we pay for an ad in the Journal of Labor Economics, and further notes that a session on point/counterpoint of the minimum wage would be a good session topic.
- *Sociologists*—Ruth Milkman is someone to reach out to for interacting with sociologists.
- *Public Policy*—Lisa Lynch at Heller School at Brandeis and Charles Jeszeck at the USGAO might be good contacts to discuss how to reach people who deal in public policy to become LERA members.
- *Labor*—AFL-CIO, IAM, Teamsters, and Steelworkers (Steve Sleigh), CWA, AFSCME, Amer Postal Workers and NJ Ed Assn (Jeff Keefe).
- *Building Construction Trade*—Training, PLA's, infrastructure investing (Joel Cutcher-Gershenfeld, Dale Belman). The board would like to recommend an award for joint labor-management recognition for both Dale Belman and Sean McGarvey to be presented at the next LERA Annual Meeting.
- *Adjunct/Grad Student Organiations*—connect at the school level. AAUP-Liebowitz (Bonnie Castrey), AAUP/AFT (Patrice Mareschal to pursue organizational membership. They meet in May/June. Patrice will speak to Jewell Gould, and Emily will contact Patrice about AFT.
- *LERA/NAA*—Joint sessions at respective meetings (regional meetings?)
- *Boeing*—John Budd
- *BCBSA NLA*—Bonnie Summers
- *Pepsi, Coke, Exxon, Shell, Ford, Campbell Soup, Disney*
- *Nurses Unions*—Jeff Keefe (AFT) Ann Tomeey, National Nurses United

Adjournment—The conference call meeting was adjourned at 3:32 a.m. by President Bonnie Castrey.

LERA Executive Board Meeting Minutes
Noon to 2 pm, Friday, May 27, 2016
Millennium Hotel, Minneapolis, MN

In attendance:

- ***Officers:*** *Bonnie Castrey (Past President), Janice Bellace (President Elect), Marty Mulloy (Past President), and Bill Canak (NCAC Chair)*
- ***Board Members:*** *John Budd, Alex Colvin, Dennis Dabney, Bill Dwyer, Rebecca Givan, Owen Herrnstadt, Tazewell Hurst III, Ken May, and Saul Rubinstein.*
- ***Committee Chairs:*** *Marlene Heyser (Development), David Lewin (Strategic), and Jeff Keefe (Membership).*
- ***Special Guests:*** *Bonnie Summers, Chris Preiner, and Tony Seguibo, of BlueCross Blue Shield Assn. NLO.*
- ***LERA Staff:*** *Emily Smith*

Not in attendance: *Craig Olson, Secrfetary-Treasurer; Eric Fidoten; Charles Jeszeck; Patrice Mareschal, and Kris Rondeau.*

Call to order. The meeting was called to order at 11:55 a.m. by Bonnie Castrey, President.

New Business

Marlene Heyser, Bonnie Summers, Tony Seguibo, and Chris Preiner presented a proposal for LERA to review the feasibility of providing educational opportunities for the staff of BlueCross BlueShield Association in several locations each year around the United States. The board approved this motion to review the feasibility of such a program, and Marlene Heyser will follow up with more details at future meetings.

Approval of the Minutes

The minutes of the January 5, 2016 LERA Executive Board Meeting and the April 15, 2016 LERA Executive Board Meeting were reviewed and unanimously approved by the Board.

Committee Reports

Financial Report—reported by Bonnie Castrey, LERA President. Craig Olson is ill in the hospital and unable to present the financial report. The financial report was reviewed mid-year 2016 figures, and accepted by the board. Mid-year income is on track with projections, although sponsorships are still unaccounted by and if not procured, will be a large impact to our bottom line. We can also expect to receive an additional $11,300 in income from the ISA at the conclusion of the 68th Annual Meeting in Minneapolis to offset the cost of their meals. In expenses, LERA overspent our projected budget for membership promotion in an effort to reach people by mail this year, as much of our online processes are in the process of being fixed, and we still have some promotions to factor in this fall. Also over budget is expense area of awards; 8 additional awards were purchased this year than budgeted for in our projections.

Editorial Committee Report—reported on by Ariel Avgar, LERA Editor-in-Chief. The LERA 2017 Research Volume proposal "No One Size Fits All: Worker Organization, Policy and Movement in a New Economic Age" edited by Janice Fine was discussed and accepted by the board. The LERA 2018 Research Volume proposal "The Contradictions of Pension Fund Capitalism" proposed by Johanna Westar was also discussed and approved by the LERA Executive Board.

Nominating Committee-reported by Bonnie Castrey. Stephen Sleigh has been picked to head up our nominating committee this year. His committee will be Dennis Dabney, Segio DelGado, Sheila Mayberry, Jodi Hoffer Gittell and Tazewell Hurst. Those who are elected will serve three-year terms on the board beginning June 1, 2017 and through May 31, 2020. The President Elect will also stand unopposed and if elected, will serve as President elect June-June 2017-18, and as President June-June in 2018-19 and Past President June-June in 2019-20.

Strategic Thinking Committee—reported on by Bonnie Castrey and David Lewin. The LERA Strategic Thinking Committee reported on a comprehensive set of bylaws revisions to be voted on in the LERA General Membership meeting, and all recommendations were accepted by the LERA board unanimously. The changes to LERA's bylaws will be distributed at the LERA General Membership meeting on Saturday, May 28, 2016 and put to the LERA membership for a vote.

LERA staff has checked whether or not meeting attendees at the LERA 69th Annual Meeting are current LERA members, and has colored their meeting registration badges accordingly. The vote will take place with people holding up their color-coded name badges to be counted, as only current LERA members can vote in the LERA General Membership meeting.

The LERA Executive Board will again meet in Chicago in conjunction with the LERA Winter Meeting on Sunday, January 8, 2017, and then in Anaheim, CA in conjunction with the LERA 69th Annual Meeting, on Friday, June 2, 2017.

Meeting was adjourned at 1:55 p.m. by Bonnie Castrey.

LERA General Membership Meeting and Awards Ceremony
5:30 p.m., May 28, 2016
North Grand Ballroom, Hotel Millennium, Minneapolis, MN

Call to order. The meeting was called to order at 5:42 p.m. by Bonnie Castrey, President.

Committee Reports

Development and Contributions Committee Report. Marlene Heyser reported that LERA will begin a new funding initiative to raise $50,000 this year. Additionally, there will be an effort to educate LERA members to how to leave a gifts from their estates.

Finance and Membership Report. Executive Director Emily Smith reported that we have made remarkable progress this past year to dig out from past years' deficit. Thanks to several sponsors, including Kaiser Permanente, and from the positive cash flow created by this year's annual meeting, we are looking to be able to get Emily help in the office. Dues for 2017 will be held at the current levels. With the new Chapter Affiliate initiative, the association is hoping for more membership growth next year as well.

Editorial Committee Report. Ariel Avgar, Editor in Chief reports that the 2017 Research Volume will be a volume edited by Johanna Westar, "The Contradictions of Pension Fund Capitalism". The Committee is currently preparing concepts for the 2018 and 2019 volumes, including Janice Fine's proposal for a volume on "No One Size Fits All: Worker Organization, Policy and Movement In a New Economic Age", which had initially been tapped as the 2017 edition, but production changes pushed it into 2018. Avgar reported that another task of the Editorial Committee is to pre-review submitted papers for the Best LERA Papers to be published in an *ILR Review* issue. The Committee thanked all who submitted papers for Competitive Papers and for the many paper sessions in the meeting.

Program Committee Report. Janice Bellace, President Elect and Program Chair for the LERA 2017 Annual Meeting in Anaheim CA, on June 1-4 at the Hilton Anaheim Hotel shared that the meeting theme is "Jobs, Opportunity and Equity in the New World of Work" and the call for proposals has been issued. Deadline is November 15, 2016.

NCAC Chapter Report and Star Award. Bill Canak, Chair of the National Chapter Advisory Committee, reported on the chapters, including a possible new startup of a Minnesota Chapter and a new student chapter in New Jersey (RU LERA). He presented a 2016 Chapter Star Award to the Long Island NY LERA Chapter.

Strategic Thinking Committee Report. David Lewin, Past President and Chair explained a new initiative from the committee and Board meant to involve chapter members in the national LERA and requiring national members to belong to a chapter if a chapter is located their area. A new "affiliate" status would be given to all chapter members allowing them access to all LERA publications, news, job board, and other services available from the LERA website for a nominal additional fee from the chapters. Additionally, the Board will be expanded to include 3 new regional vice presidents from the chapters, alternating regions and perspectives. The motion was made

> "That the language in the LERA Bylaws and Constitution be amended to reflect new affiliate status of LERA Chapter members, new requirements of the chapters to work with the LERA, the addition of three regional vice presidents to the LERA Board to represent Chapters, and other minor changes to the grammar and syntax of the documents."

David Lewin, presenting the motion on behalf of the committee, noted that LERA's current bylaws permit changes to be made by vote at the General Membership Meeting, and that only national members in good standing were eligible to vote (said members being signified by the color of their name card). The motion to amend the bylaws was seconded and discussion ensued. All of the discussion related to the issue of the new category of affiliated membership for local chapter members, who would pay nominal dues. Rudy Olswald of LERA Gateway Chapter (St. Louis, MO) stated that he did not believe individual local chapters had sufficient time in which to discuss the change regarding affiliate members. Bill Canak, NCAC chair, explained how local chapter representatives had been informed. An amendment to the motion was then made; namely, to have the vote on the bylaws changes be done via mail ballot. After more discussion, the question was called. The amendment to the original motion was defeated, by a vote of 24 to 18. Then a vote was taken on the original motion. The vote was 41 in favor and 3 against, and the motion to amend the bylaws as submitted by the bylaws revision committee was approved. President Bonnie Castrey noted that a bylaws implementation committee would established as there were many implementation issues to be resolved.

2016 Awards Ceremony

Best Dissertation Award Committee. Bruce Kaufman of the Thomas A. Kochan and Steven R. Sleigh Best Dissertation Award Committee asked President Bonnie Castrey to award the Best Dissertation Award to co-winners: Mark Gough, Cornell University and Chunyun Li, Rutgers University. Two honorable mentions were also awarded to Qianyun Xie, University of Minnesota and Sarah Young, University of Tennessee-Knoxville

LERA Awards Committee. Co-Chairs Doug Kruse and Tia Denenberg presented the 2016 slate of awards.

- The John T. Dunlop Outstanding Scholar Awards were given to co-winners John-Paul Ferguson, Stanford University and Aaron Sojourner, University of Minnesota for national research issues and the winner for international contributions was Xiangmin "Helen" Liu, Pennsylvania State University.

- The Outstanding Practitioner Awardee was Allison Beck, FMCS. Accepting for Allison was Scot Beckenbaugh.

- The Susan C. Eaton Outstanding Scholar-Practitioner Award went to David Weil, the US Department of Labor and Boston University.

- Academic Fellows for 2016 are Morris Kleiner, University of Minnesota; Roy Adams, McMasters University; Teresa Ghilarducci, New School for Social Research; and Lisa Blomgren Amsler, Indiana University-Bloomington. Practitioner Fellows are Marcia Greenbaum, Workplace Solutions; Marlene Heyser, Workplace Law Strategies; and Eric Mooshegian, International Union of Operating Engineers.

James Scoville Best International Paper Committee. John Budd presented the 2016 Best International Paper Award to co-winners: Valeria Pulignano, K.U. Leuven and Jasmine Kerrissey, University of Massachusetts-Amherst.

New and Other Business

The next LERA Annual Meeting will be held June 1-4, 2017 in Anaheim CA, at the Anaheim Hilton Hotel. After asking for any New or Other Business items and hearing none, President Bonnie Castrey thanked the Executive Board and all present and handed over the gavel to Janice Bellace, the new LERA President. President Bellace adjourned the meeting at 7:25 p.m.

LERA Executive Board Meeting Minutes
Monday, December 12, 2016, 2 p.m. ET
Via Conference Call

Call to order—the meeting was called to order at 2:00 p.m. by Janice Bellace, LERA President. Present at the meeting were officers Janice Bellace, President, Bonnie Castrey, Past President, Harry Katz (President-Elect); Bill Canak, NCAC Chair; Craig Olson, Secretary-Treasurer; Ariel Avgar, Editor-in-Chief. Board Members present at the meeting were Annette Bernhardt, John Budd, Jonathan Donehower, Eric Fidoten, Rebecca Givan, Owen Herrnstadt, Charles Jeszeck, Patrice Mareschal, Dan Marschall, Kris Rondeau, Saul Rubinstein, Jeff Wheeler. Committee Chairs present included Marlene Heyser, Jeff Keefe, and Steve Sleigh. Emily Smith was present from LERA staff. Not attending the meeting were Dennis Dabney and Beth Schindler.

Reports

Approval of the Minutes—President Janice Bellace presented the minutes from the General Membership Meeting in Minneapolis. They were approved unanimously with no abstentions and no no's.

Nominating Committee Report—Secretary-Treasurer Craig Olson reports on the slate from this year's Nominating Committee. Steve Sleigh reported that the committee met in November and sought to find quality nominees with balance on the slate in terms of gender, geographic distribution, Chapter members, and other areas of diversity. Bonnie Castrey seconded, and the board unanimously approved the slate of candidates for the LERA 2017 election with no abstention's and no no's.

This slate contains a new seat of an academic board member that is reserved specifically for an academic that represents a school that does not have a dedicated IR/HR program, or that is a small school, or someone who is a lecturer and not a tenure-track professor, etc. This first seat will have a two-year term so that we will have 5 candidates that expire each year, but will continue with a three-year term following that.

There will be forthcoming nominees and alternates to run for the new position of Regional Vice President (of which there will be three, one to represent Chapters in each region: East, West, and Mid). These names will come from a sub-committee of NCAC members and Chapter leaders and we will add these to the election in accordance with our bylaws revisions that were voted on in Minneapolis. Bonnie Castrey can elaborate on this process when she talks about Implementation Committee business.

Program Committee Reports—Janice Bellace reports on the progress of both the LERA Winter Meeting 2017 and the LERA 69th Annual Meeting in Anaheim, CA. For the LERA 69th Annual Meeting we had about 55 session submission proposals and the session looks full and robust. The Program Committee is currently checking for obvious gaps in programming and there is a good balance between practitioner and academic sessions, and that we have good content from the West Coast and examining plenary sessions. By the time LERA meets in Chicago, the program should be more or less finalized. Most of the comments from the program committee had to do with having more descriptive titles. Becky Givan wants to emphasize that the plenary sessions should consider gender, race, age, and constituency diversity in our plenary speakers.

The LERA 2017 Winter Meeting will take place at the Hyatt Regency in January 6-8, 2017. Sandy Jacoby and Jeannette Wicks-Lim have headed this up, and this is Sandy's last year. He'll be replaced by Susan Johnson at Upjohn next year, and the program looks terrific. The board members are encouraged to attend, and the next LERA Board meeting will take place on site. Emily will staff an information booth, and please RSVP if you plan to attend the board meeting. The LERA 2018 Winter Meeting will take place in Philadelphia.

Implementation Committee Report—Bonnie Castrey reports. The committee has met several times, and we have held two conference calls asking Chapters to participate, of which 17 have participated. The NCAC will continue to work with the Implementation Committee to continue to get participation from the remaining chapters. The committee is recommending that all the Chapters east of Ohio (Ohio and East) be in the East Region, all the Chapters Colorado and West be in the West region, and everything in between be considered the Mid region. That approximates roughly equal numbers of affiliates for each region, as chapters self-report members. Affiliates will not be members of national, but they will receive our e-Bulletins. The other large area that we talked about in rolling this out is setting up three nominating sub-committees so that people who really know people in those regions can nominate a couple of good folks from each region, and get those people on the election ballot. Joel Cutcher-Gershenfeld is working to put out a survey to all of the chapters and Bill Canak will continue to work on participation. The dues structure for Chapter to national dues has not been reviewed since 2013, and the Implementation Committee recommends that the board review the dues structure as a separate process from the implementation committee process. The doubling of dues is seen as recent history, not ancient history by Chapters, and caution is advised when reviewing dues changes.

National Chapter Advisory Committee—William Canak reports. Penn State University, which has an extensive online program, is developing with the NCAC an online chapter for its online students. Also, more of our chapters are being more aggressive on offering MCLE/CLE and SHRM credits for their workshops. The NCAC will continue to work with chapters as the implementation process proceeds. In Anaheim, the Chapter Reps will meet, and for the second year, the Chapters will be making contributions to sponsor and host a Chapter reception in Anaheim. There will also be a workshop on various aspects of chapter administration, on finances, administration, programming, etc.

Financial Report, Audit and Budget—Craig Olson reports. The audit has been conducted for 2015 successfully with two findings, first the auditor suggests that LERA have two signatories on checks written, and second, that staff should keep track of hours by program area as a cost accounting measure. The budget is tracking towards being in the black for 2016, due to a successful annual meeting, keep expenses low, and fixing the MemberClicks functionalities.

Biggest budget consideration for 2016 and 2017 is adding second full-time staff member. LERA has operated with one full-time staff member since August 2015. Bernadette Tiemann has been working part-time (20 hours per week as a temporary worker) for about six months; and the current plan is to move forward with hiring a full-time permanent staff member by January 16, 2017.

Editorial Committee Report—Ariel Avgar reports that the 2016 edition is just about to hit mailboxes, and that the 2017 edition, edited by Johanna Westar is asking for draft chapters by January and completed chapters by June 2017. We also have a 2018 volume lined up edited by Janice Fine on experimentation and low-wage worker organization. They are in early stages of establishing chapters and authors, but should be in line for 2018. The Editorial Committee is pleased to have two volumes out and is hopeful to solicit proposals to submit to the board in June, possibly interested in a topic on education, and if anyone has any ideas, please forward them to the committee. The committee has not traditionally sent out a CFP, they have informally solicited proposals. Janice Bellace and Marlene Heyser suggest LERA send out a request for proposals on topics for Research Volumes to Chapters in January 2017.

The LERA Best Papers issue of the Cornell ILR Review is taking place now from the Minneapolis meeting. They received 10 papers from the meeting, and 5 were taken for consideration, and we are hoping to have 2 or 3 that make it into the journal, and we hope that this process can be well marketed for the Anaheim meeting so that authors have an understanding of the process and the opportunity that is available. The goals of this were two-fold. One, to have an institutional relationship with an academic journal, and secondly, often younger scholars want to see a clear linkage between opportunities to present research and then to have it published it in a scholarly journal. The five papers that got sent through were of high quality, and hopefully if we have enough high quality submissions in the future, we are hopeful to get even a special issue in the future.

Development Committee Report—Marlene Heyser Committee chair reports. When Eric was the Executive Director, he made the contacts himself with Ford, and he didn't communicate well to the Development Committee, so it's possible that we've not been able to. John Flemming sits on a board with Eric Fidoten and Eric may be able to make some in-roads there. The board recommends that we re-establish communication with Marty Mulloy, Bill Dirksen, and Jimmy Settles, and that Eric Fidoten contact John Flemming. Janice Bellace suggests that wherever this budget is coming from has a contact individual, and we need to find who that person is at the Ford-UAW partnership committee.

David Lewin and Marlene Heyser are continuing to explore the relationship between BlueCross BlueShield Association National Labor Office to establish educational opportunities for their staff on an annual basis utilizing LERA educators with the idea that it could represent a revenue stream for LERA on a yearly basis. They are thinking of an East Coast and a West Coast program, and perhaps even Midwest program in a a few years. For OCLERA, they have dinner meetings every other month. On May 30, 2017, the OC LERA will conduct its monthly meeting that evening, very close to the Hilton where LERA will meet. They will invite LERA national meeting attendees, and also promote the LERA 69th Annual Meeting to their own members. Jim Pruitt is the co-chair for this event for OC LERA.

Membership Committee Report—Jeffrey Keefe reports that membership is low but appears to have possibly bottomed-out. The areas where member numbers have dwindled are Emeritus and Regular members. Last year's effort was to attempt to bring in new members from the Chapters, and we might be in the very start of seeing some results from that effort.

What types of content and services would be appealing to newly organizing student worker unions? If we are going to invite them to join the organization, then we should have content and services that appear to them. Board suggests that we invite them in to decide what the programming should be in Anaheim to meet their needs, and that we discuss this further in January 2017 at our next meeting.

The last election could present a crisis for labor in this country and may result in the opportunity to receive grant funds to address the crisis. Much of what has been said about the Trump administration is contrary to LERA's values. LERA could play the role or promoting serious and thoughtful bi-partisan dialogue from people from different perspectives. Let's identify entities who would fund research projects in this vein and discuss further in January.

Recruiting student members into LERA is to use the Wall Street Journal model of recruiting ten students into joining LERA, and that professor would receive a free annual membership in return. The students at Rutgers run their own Chapter, and they seem to be very interested in what the bigger questions about labor and employment relations that students are interested. Students want networks and jobs, so let's give more thought on how we can better serve them.

Bill Canak has suggested that LERA members should receive a membership card annually in the mail to keep in their wallet; let's review this idea again in Chicago and get feedback from the board and consider the cost and benefits and how to best implement the idea if we wish to proceed. Staff should try to bring samples and prices to the January meeting, along with an example of the Card as suggested by Bill.

New Business

LERA Hotel Contract in Baltimore 2018—LERA reviewed location options for 2018, in the East region, and gave consideration to union hotels and to locations of LERA Chapters. Janice Bellace reached out to the Maryland chapter and they were very excited to act as a host Chapter for 2018 and organizing efforts are already underway. Baltimore is a nice blue-collar contract to Washington, DC, and the local chapter seems to have good ties to the city administration.

Cleveland is on our short list for 2019, and they do have a beautiful new union hotel, so this is looking like a good choice. Dan Marschall also supports the choice of Cleveland for 2019.

Webcasting and LERA events—LERA can explore the option of obtaining a webcasting service for a fixed fee annually. LERA has obtained a thirty-day free trial, thanks to a suggestion from Harry Katz, and we will use the thirty-day free trial to conduct webcasting at the LERA 2017 Winter Meeting in Chicago, in preparation for the Anaheim meeting. This service, once paid for, provides two concurrent licenses (web casting rooms) to webcast two events at any given time, so LERA Chapters and other LERA working groups could tap into this service if this is helpful and valuable. It could add value to chapters, and LERA could be developing a video catalogue that would also represent a value to LERA members.

Adjournment—The conference call meeting was adjourned at 2:54 p.m. by President Janice Bellace. The motion was unanimously passed by the board.

Labor and Employment Relations Association

Champaign, Illinois

Financial Statements

For the Years Ended
December 31, 2015 and 2014

Feller & Kuester CPAs LLP
Certified Public Accountants
806 Parkland Court, Suite 1
Champaign, IL 61821
217-351-3192

TABLE OF CONTENTS

Feller & Kuester CPAs LLP

Tax - Audit - Bookkeeping
806 Parkland Court, Suite #1 - Champaign, Illinois 61821
Phone - (217) 351-3192 Fax - (217) 351-4135 Email - neal@fellerkuester.com

INDEPENDENT AUDITOR'S REPORT

To the Board of Directors of
Labor and Employment Relations Association
Champaign, Illinois

We have audited the accompanying financial statements of the Labor and Employment Relations Association (a nonprofit organization) which comprise the statements of financial position as of December 31, 2015 and 2014, and the related statements of activities and cash flows for the years then ended, and the related notes to the financial statements.

Management's Responsibility for the Financial Statements
Management is responsible for the preparation and fair presentation of these financial statements in accordance with accounting principles generally accepted in the United States of America; this includes the design, implementation, and maintenance of internal control relevant to the preparation and fair presentation of financial statements that are free from material misstatement, whether due to fraud or error.

Auditor's Responsibility
Our responsibility is to express an opinion on these financial statements based on our audits. We conducted our audits in accordance with auditing standards generally accepted in the United States of America. Those standards require that we plan and perform the audit to obtain reasonable assurance about whether the financial statements are free from material misstatement.

An audit involves performing procedures to obtain audit evidence about the amounts and disclosures in the financial statements. The procedures selected depend on the auditor's judgment, including the assessment of the risks of material misstatement of the financial statements, whether due to fraud or error. In making those risk assessments, the auditor considers internal control relevant to the entity's preparation and fair presentation of the financial statements in order to design audit procedures that are appropriate in the circumstances, but not for the purposes of expressing an opinion on the effectiveness of the entity's internal control. Accordingly, we express no such opinion. An audit also includes evaluating the appropriateness of accounting policies used and the reasonableness of significant accounting estimates made by management, as well as evaluating the overall presentation of the financial statements.

We believe that the audit evidence we have obtained is sufficient and appropriate to provide a basis for our audit opinion.

Opinion

In our opinion, the financial statements referred to above present fairly, in all material respects, the financial position of the Labor and Employment Relations Association as of December 31, 2015 and 2014, and the changes in its net assets and its cash flows for the years then ended in accordance with the accounting principles generally accepted in the United States of America.

Other Matter

Our audit was conducted for the purpose of forming an opinion on the financial statements as a whole. The schedules of functional expenses on pages 16 and 17 are presented for purposes of additional analysis and are not a required part of the financial statements. Such information is the responsibility of management and was derived from and relates directly to the underlying accounting and other records used to prepare the financial statements. The information has been subjected to the auditing procedures applied in the audit of the financial statements and certain additional procedures, including comparing and reconciling such information directly to the underlying accounting and other records used to prepare the financial statements or to the financial statements themselves, and other additional procedures in accordance with auditing standards generally accepted in the United States of America. In our opinion, the information is fairly stated in all material respects in relation to the financial statements as a whole.

Feller & Kuester CPAs LLP

Feller & Kuester CPAs LLP
Champaign, Illinois

August 23, 2016

LABOR AND EMPLOYMENT RELATIONS ASSOCIATION
STATEMENTS OF FINANCIAL POSITION
DECEMBER 31, 2015 AND 2014

	2015	2014
ASSETS		
Current Assets		
Cash and Cash Equivalents	$ 136,263	$ 226,823
Investments	23,145	22,856
Accounts Receivable, Net	911	250
Prepaid Expenses	2,296	6,256
Inventory	6,729	5,537
Accrued Interest and Royalties	2,617	4,305
Total Current Assets	171,961	266,027
Property and Equipment, Net	848	1,587
TOTAL ASSETS	$ 172,809	$ 267,614
LIABILITIES AND NET ASSETS		
Current Liabilities		
Accounts Payable	$ 575	$ 8,085
Funds Held for Other Organizations	5,805	7,516
Accrued Liabilities	15,557	24,315
Dues Collected in Advance	54,742	56,757
Subscriptions Collected in Advance	10,484	7,278
Deferred Sponsorships	-	25,000
Deferred Chapter Dues	6,300	8,560
Other Deferred Revenue	-	1,755
Total Current Liabilities	93,463	139,266
Net Assets		
Temporarily Restricted		
Eaton Memorial Scholarship Receipts	32,040	31,935
Kochan-Sleigh Award Receipts	43,090	43,624
Scoville Award Receipts	6,455	7,041
Dunlop Fund	19,809	19,474
Gershenfeld Fund	23,298	23,030
Total Temporarily Restricted	124,692	125,104
Unrestricted Net Assets	(45,346)	3,244
Total Net Assets	79,346	128,348
TOTAL LIABILITIES AND NET ASSETS	$ 172,809	$ 267,614

See Accompanying Notes.

LABOR AND EMPLOYMENT RELATIONS ASSOCIATION
STATEMENTS OF ACTIVITIES
FOR THE YEARS ENDED DECEMBER 31, 2015 AND 2014

	2015			2014		
	Unrestricted	**Temporarily Restricted**	**Total**	**Unrestricted**	**Temporarily Restricted**	**Total**
Support and Revenues						
Membership Dues	118,190	-	118,190	137,775	-	137,775
Meeting Income	87,484	-	87,484	105,080	-	105,080
Organization Dues & Sponsorships	49,000	-	49,000	70,800	-	70,800
Subscriptions	18,546	-	18,546	20,005	-	20,005
Chapter Fees	12,760	-	12,760	11,300	-	11,300
Member Contributions	11,499	315	11,814	9,772	6,422	16,194
Royalties	5,772	-	5,772	11,587	-	11,587
Ad Income	1,400	-	1,400	1,000	-	1,000
Publications	257	-	257	347	-	347
Mailing List Rental	238	-	238	723	-	723
Administrative Fees	98	-	98	834	-	834
Investment Income	-	289	289	-	1,483	1,483
Interest Income	314	177	491	345	160	505
Net Assets Released from Restrictions	1,193	(1,193)	-	4,205	(4,205)	-
Total Support and Revenues	306,751	(412)	306,339	373,773	3,860	377,633
Expenses						
Program Services						
General	166,237	-	166,237	187,521	-	187,521
Meetings	89,809	-	89,809	78,968	-	78,968
Publications	41,604	-	41,604	41,541	-	41,541
Supporting Services						
Management and General	50,979	-	50,979	80,714	-	80,714
Membership Development	6,712	-	6,712	10,691	-	10,691
Total Expenses	355,341	-	355,341	399,435	-	399,435
Change in Net Assets	(48,590)	(412)	(49,002)	(25,662)	3,860	(21,802)
Net Assets, Beginning of Year	3,244	125,104	128,348	28,906	121,244	150,150
Net Assets, End of Year	$ (45,346)	$ 124,692	$ 79,346	$ 3,244	$ 125,104	$ 128,348

See Accompanying Notes.

LABOR AND EMPLOYMENT RELATIONS ASSOCIATION
STATEMENTS OF CASH FLOWS
FOR THE YEARS ENDED DECEMBER 31, 2015 AND 2014

	2015	2014
CASH FLOWS FROM OPERATING ACTIVITIES		
Change in Net Assets	$ (49,002)	$ (21,802)
Adjustments to Reconcile Change in Net Assets to Net Cash Provided by (Used in) Operating Activities		
Depreciation Expense	739	1,080
Loss on Disposal of Fixed Assets	-	445
Net Unrealized (Gains) Losses on Investments	412	(715)
(Increase) Decrease in Operating Assets		
Accounts Receivable, Net	(661)	26,306
Prepaid Expenses	3,960	481
Inventory	(1,192)	1,060
Accrued Interest and Royalties	1,688	(659)
Increase (Decrease) in Operating Liabilities		
Accounts Payable	(7,510)	(8,004)
Funds Held for Other Organizations	(1,711)	2,581
Accrued Liabilities	(8,758)	2,256
Dues Collected in Advance	(2,015)	(22,319)
Subscriptions Collected in Advance	3,206	(3,018)
Deferred Sponsorships	(25,000)	25,000
Deferred Chapter Dues	(2,260)	860
Other Deferred Revenue	(1,755)	855
Net Cash Provided by (Used in) Operating Activities	(89,859)	4,407
CASH FLOWS FROM INVESTMENT ACTIVITIES		
Reinvested Interest, Dividends, and Capital Gains	(701)	(768)
Net Cash Provided by (Used in) Investment Activities	(701)	(768)
CASH FLOWS FROM FINANCING ACTIVITIES		
None	-	-
Net Cash Provided by (Used in) Financing Activities	-	-
Net Increase (Decrease) in Cash and Cash Equivalents	(90,560)	3,639
Beginning Cash and Cash Equivalents	226,823	223,184
Ending Cash and Cash Equivalents	$ 136,263	$ 226,823

See Accompanying Notes.

LABOR AND EMPLOYMENT RELATIONS ASSOCIATION
NOTES TO FINANCIAL STATEMENTS
DECEMBER 31, 2015 AND 2014

NOTE 1 – NATURE OF ACTIVITIES AND SIGNIFICANT ACCOUNTING POLICIES

Organization and Nature of Activities

The Labor and Employment Relations Association (the Organization) was founded in 1947 to encourage research in all aspects of the field of labor, employment, and the workplace. It is a nonprofit scholarly association of academic, labor, business, and neutral communities committed to the full discussion and exchange of ideas between and among its broad constituencies through meetings, publications, and its various electronic listservs and websites. The Organization's National Office is located in Champaign, Illinois and serves the association by planning conferences and meetings and publishing the various researches of its members. The main sources of support and revenues for the Organization are contributions, membership dues and subscriptions, and meeting income.

Subsequent Events

Subsequent events have been evaluated through August 23, 2016, which is the date the financial statements were available to be issued.

Accrual Basis of Reporting

The Organization has chosen to report on the accrual basis of accounting. Accordingly, revenue is recognized when earned and expenses are recognized when incurred in conformity with accounting principles generally accepted in the United States of America (GAAP).

Estimates

The preparation of financial statements in conformity with GAAP requires management to make estimates and assumptions that affect certain reported amounts and disclosures. Accordingly, actual results could differ from those estimates.

Cash and Cash Equivalents

For purposes of reporting cash flows, cash and cash equivalents include all cash and highly liquid investments acquired with an original maturity date of three months or less. Since the penalties of converting certificate of deposits to cash is insignificant, all certificate of deposits have been included with cash and cash equivalents. As of December 31, 2015 and 2014, \$101,547 and \$102,248 of cash and cash equivalents are temporarily restricted and are subject to donor-imposed stipulations.

Investments

Investments consist of intermediate term bond index funds. All investments are held as temporarily restricted and are subject to donor-imposed stipulations.

Accounts Receivable

Accounts receivable are recorded primarily for outstanding invoices for membership dues and subscriptions. An allowance for doubtful accounts is based on an analysis of expected collection rates determined from experience. The Organization had an allowance for doubtful accounts of $100 at December 31, 2015 and 2014.

Inventory

The Organization's inventory of directories, research volumes, proceedings, and perspective magazines is carried at the lower of cost and market value. Cost is determined on the basis of first in – first out.

Property and Equipment

Property and equipment expenditures in excess of $500 are capitalized at cost. Donated property and equipment are capitalized at estimated cost or fair market value at the time of donation. Depreciation of the assets is computed using the straight-line method over their estimated useful lives. The range of estimated useful lives by type of asset is as follows:

Furniture and Equipment	5-7 years

Net Assets

Net assets of the Organization and changes therein are classified and reported as follows:

Unrestricted Net Assets – Net assets that are not subject to donor-imposed stipulations.

Temporarily Restricted Net Assets – Net assets subject to donor-imposed stipulations that may or will be met, either by actions of the Organization and/or passage of time. When a restriction expires, temporarily restricted net assets are reclassified from restrictions. Contributions received with temporarily restrictions that are met in the same reporting period are reported as unrestricted support and increase unrestricted net assets. The Organization had $124,692 and $125,104 of temporarily restricted net assets as of December 31, 2015 and 2014, respectively.

Permanently Restricted Net Assets – Net assets subject to donor-imposed stipulations that they be maintained permanently by the Organization. Generally, the donors of these assets permit the Organization to use all or part of the income earned on any related investments for general or specific purposes. The Organization had no permanently restricted net assets as of December 31, 2015 and 2014.

Grant Revenue

Grant funds received under contracts that require provision of services by the Organization are recognized as revenue to the extent that expenses have been incurred for the purpose specified by the grantor. The excess of grant monies received over expenses incurred is accounted for as deferred revenue. The excess of grant expenditures over grant monies received is accounted for as grant revenue receivable to the extent that additional grant revenue has been awarded and will be paid.

Grant funds received with only general restrictions and public donations are considered to be available for unrestricted use unless specifically restricted by the donor. When such restrictions exist, the revenue is recorded as temporarily restricted until the restrictions are met.

Contributed Services and Goods

Contributed services are reported as contribution revenue and as assets or expenses only if the services create or enhance a non-financial asset (for example property and equipment) or:

- Would typically need to be purchased by the Organization if the services had not been provided by contribution.
- Require specialized skills.
- Are provided by individuals with those skills (such as accounting, financial, construction, educational, electrical, legal, medical, and other services provided by accountants, investments advisers, contractors, teachers, electricians, lawyers, doctors, and other professional and craftspeople).

For the years ended December 31, 2015 and 2014, the value of contributed services meeting the defined requirements for recognition in the financial statements as outlined above were not material and have not been recorded on the financial statements. The Organization does receive free office space from the University of Illinois. However, the value for the use of this office space has not been reported on the financial statements.

Income Tax Status

The Organization is a nonprofit association that is exempt from federal income tax under Section 501 (c)(3) of the Internal Revenue Code. In addition, the Internal Revenue Service has determined that the Organization is not a private foundation as defined in Section 509(a)(1) and Section 170(b)(1)(A)(vi) of the Code.

The Organization has evaluated its exposure resulting from uncertain income tax position and determined the exposure is not material to the financial statements. In addition, the Organization is not aware of any tax position for which a significant change is reasonably possible within the next 12 months. Therefore, these financial statements do not include a liability for uncertain tax positions. Upon recognition of a liability for an uncertain tax position, the Organization would recognize interest expense and penalties in operating expenses.

The Organization files information tax returns in the U.S. federal jurisdiction and the state of Illinois. Its federal and Illinois information tax returns prior to fiscal year 2012 are closed. The Organization does not have any tax returns currently under examination by either the Internal Revenue Service (IRS) or any U.S. state jurisdiction.

Membership Dues and Advance Subscriptions Collected

Membership dues and subscriptions are assessed and recognized as revenue based on the life of the dues or subscription.

Functional Allocation of Expenses

The costs of providing the various programs and other activities have been summarized on a functional basis in the statement of activities. Accordingly, certain costs have been allocated among the programs and supporting services benefited.

NOTE 2 – ARRANGEMENTS WITH THE UNIVERSITY OF ILLINOIS

The Organization moved its offices to the University of Illinois at the end of 1999. Under an arrangement with the University, the employees of the Organization are employed by the University. The employees' pension and benefits are part of the University's plans. The Organization then reimburses the University quarterly for the cost of its employees.

The University of Illinois holds some cash for the Organization. These "claim on cash" balances were $12,632 and $21,614 as of December 31, 2015 and 2014, respectively. These balances have been included with cash and cash equivalents.

NOTE 3 – UCIRHRP SERVICING ARRANGEMENT TO ADMINISTER BUSINESS

On January 3, 2003, the Organization entered into a servicing arrangement to administer the business of the University Council of Industrial Relations and Human Resources Programs (UCIRHRP). These servicing duties include dues notification and collection, annual meeting arrangement and report preparation, and maintaining a data base and network communications for this separate organization. A separate financial statement is maintained for this company, but UCIRHRP's cash is maintained in the Organization's general bank account. The amount of cash belonging to UCIRHRP in the general account is shown on the Organization's balance sheet as Funds Held for Other Organizations. UCIRHRP's cash balance in the Organization's custody was $5,805 at December 31, 2015 and $5,865 at December 31, 2014. The Organization is allowed to keep 20% of dues collected each year as an administrative fee.

NOTE 4 – CONCENTRATIONS OF CREDIT RISK

Financial instruments that potentially subject the Organization to credit risk consist principally of checking accounts, money markets accounts, and certificates of deposits at financial institutions. However, management continuously monitors the Organization's balances at financial institutions. Deposits held at any financial institutions were fully insured by the Federal Deposit Insurance Corporation (FDIC) at December 31, 2015 or 2014. The "claims on cash" held with the University of Illinois as stated in Note 2 is not insured since it is not held in a financial institution. In addition, the Organization had undeposited funds of $0 and $7,850 at December 31, 2015 and 2014, respectively. Since these funds were not yet deposited with a financial institution it was covered by FDIC.

NOTE 5 – TEMPORARILY RESTRICTED NET ASSETS

Susan C. Eaton Scholar-Practitioner Memorial Fund

The Organization set up a memorial fund in honor of an author of a 1998 "Perspectives on Work" article and active young the Organization member, who died on December 30, 2003. At its June 1, 2004 meeting, the Organization approved the establishment of an annual Susan C. Eaton Scholar-Practitioner award and grant to be paid to one or more qualified scholar researchers in even-numbered years or practitioners in odd-numbered years doing research in the labor and employment relations or related field. The Organization's Executive Board directed that 10% of any gifts received each year are unrestricted and can be used for administrative expenses. As of December 31, 2015 and 2014, these temporary restricted funds totaled $32,040 and $31,935, respectively.

Kochan-Sleigh Best Dissertation Award Fund

On March 27, 2006, the Organization set up a temporarily restricted fund with contributions from a member and matching contributions from General Electric. The fund is restricted for a minimum of fifteen years and may be used to pay for the $1,000 annual best dissertation award and plaque. After the fifteen year period, the fund may be continued or the amount remaining in the fund may be released and become available to the Organization for unrestricted purposes. The Organization's Executive Board directed that 10% of any gifts received each year are unrestricted and can be used for administrative expenses. As of December 31, 2015 and 2014, these temporary restricted funds totaled $43,090 and $43,624, respectively.

Jim Scoville Award Fund

On January 9, 2009, the Organization set up a temporarily restricted fund with a $10,000 contribution from the University of Minnesota's Industrial Relations Center to honor a member and retiring professor. The fund was established to pay an annual award of $500 and a plaque for best paper on international and comparative employment issues. The Organization's Executive Board directed that 10% of any gifts received each year are unrestricted and can be used for administrative expenses. As of December 31, 2015 and 2014, these temporarily restricted funds totaled $6,455 and $7,041, respectively.

John T. Dunlop Public Policy Fund

On May 8, 2010, the Organization established a temporarily restricted fund to subsidize a named John T. Dunlop Public Policy Session at future the Organization National Policy Forums and/or at the Organization annual meetings. The Organization's Executive Board directed that 10% of any gifts received each year are unrestricted and can be used for administrative expenses. This fund will be so named for five or ten years, at which time the Organization's Executive Board will review the fund to determine whether to continue or modify the fund. As of December 31, 2015 and 2014, these temporary restricted funds totaled $19,809 and $19,474, respectively.

Gladys and Walter Gershenfeld Publication Fund

On May 8, 2010, the Organization established a temporarily restricted fund to honor the memory of long-time members and past presidents Walter and Gladys Gershenfeld for the purpose of supporting the Organization's electronic and print publications. The fund will be so named for ten years, at which time the Organization will review the fund's purpose and uses, and whether to continue or modify the fund. The Organization's Executive Board directed that 10% of any gifts received each year are unrestricted and can be used for administrative expenses. As of December 31, 2015 and 2014, these temporarily restricted funds totaled $23,298 and $23,030, respectively.

NOTE 6 – INVESTMENTS

Financial Accounting Standards Board ("FASB") Codification 820, Fair Value Measurements, establishes a framework for measuring fair value. That framework provides a fair value hierarchy that prioritizes the inputs to valuation techniques used to measure fair value. The hierarch gives the highest priority to unadjusted quoted prices in active markets for identical assets or liabilities (Level 1 measurements) and the lowest priority to unobservable inputs (Level 3 measurements). The three levels of the fair value hierarchy under FASB Codification 820 are described below:

Level 1 – Inputs to the valuation methodology are based on unadjusted quoted prices for identical assets or liabilities in active markets that the Organization has the ability to access.

Level 2 – Inputs to the valuation methodology include quoted prices for similar assets or liabilities in active markets, quoted prices for identical or similar assets or liabilities in inactive markets, inputs other than quoted prices that are observable for the asset or liability, and inputs that are derived principally from or corroborated by observable market data by correlation or other means. If the asset or liability has a specified (contractual) term, the Level 2 input must be observable for substantially the full term of the asset or liability.

Level 3 – Inputs to the valuation methodology are unobservable and significant to the fair value measurement.

The asset's and liability's fair value measurement level within the fair value hierarchy is based on the lowest level of any input that is significant to the fair value measurement. Valuation techniques used need to maximize the use of observable inputs and minimize the use of unobservable inputs.

The following is a description of the valuation methodologies used for assets measured at fair value. There have been no changes in the methodologies used at December 31, 2015 and 2014.

Intermediate Term Bond Funds: Valued at the NAV of shares held by the Organization at year-end based on readily determinable fair values, which are published daily and are the basis for current transactions.

The preceding methods described may produce a fair value calculation that may not be indicative of net realizable value or reflective of future fair values. Furthermore, although the Organization believes its valuation methods are appropriate and consistent with other market participants, the use of different methodologies or assumptions to determine the fair value of certain financial instruments could result in different fair value measurements at the reporting date.

The following table sets forth by level, within the fair value hierarchy, the Organization's assets at fair value as of December 31, 2015 and 2014:

Assets at Fair Value as of December 31, 2015

	Level 1	Level 2	Level 3	Total
Intermediate Term Bond Funds	$23,145	-	-	$23,145

Assets at Fair Value as of December 31, 2014

	Level 1	Level 2	Level 3	Total
Intermediate Term Bond Funds	$22,856	-	-	$22,856

Net investment earnings for the years ended December 31, 2015 and 2014 are summarized as follows:

	2015	2014
Dividends	$ 622	$ 636
Capital Gains Distributions	79	132
Net Unrealized Gains (Losses)	(412)	715
Total Investment Income	$ 289	$ 1,483

This investment income, including unrealized gains and losses, are being reported on the statement of activity as investment income.

NOTE 7 – PROPERTY AND EQUIPMENT

As of December 31, 2015, Property and Equipment consists of:

Furniture and Equipment	$ 14,551
Less: Accumulated Depreciation	(13,703)
Property and Equipment, Net	$ 848

As of December 31, 2014, Property and Equipment consists of:

Furniture and Equipment	$ 15,639
Less: Accumulated Depreciation	(14,052)
Property and Equipment, Net	$ 1,587

Depreciation expense for the years ended December 31, 2015 and 2014 was $739 and $1,080, respectively.

NOTE 8 – ACCRUED EXPENSES

As of December 31, 2015, Accrued Expenses consist of:

Accrued Payroll	$ 4,073
Accrued Benefits/Taxes	1,352
Accrued Vacation	10,132
Total	$ 15,557

As of December 31, 2014, Accrued Expenses consist of:

Accrued Payroll	$ 5,082
Accrued Benefits/Taxes	2,002
Accrued Vacation	17,231
Total	$ 24,315

NOTE 9 – COMMITMENTS

The Organization had a contract with University of Illinois Press (UIP) for UIP to provide membership management services and publication of *Perspectives on Work* (magazine) for an annual fee of $28,280. The five year agreement commenced on January 1, 2007 and expired on December 31, 2012. Although the original contract expired, the terms continued through December 31, 2014. As of December 31, 2014, the Organization is no longer under contract with UIP and no future payments are due to UIP.

On February 26, 2013, the Organization signed a 12 month contract from March 1, 2013 to February 28, 2014 with a company to provide services to manage the EPRN website and provide content for the Organization's website. The contract states that the Organization will pay this company $2,400 per month for a total of $28,800. The Agreement automatically renews for consecutive 12 month terms, unless the Agreement is terminated by one or both of the parties as set forth herein. Beginning the second cycle and continuing until termination or renegotiation, the Organization will increase the base rate every year by two percent (2%) over the previous year's base rate. For example: $2,448 per month for twelve months starting March 1, 2014; $2,497 per month for twelve months starting March 1, 2015; $2,546 per month for twelve months starting March 1, 2016; and so on.

The Organization has signed contracts with hotels in Pittsburgh, Minneapolis, and Anaheim for the 2015, 2016, and 2016 Annual Meetings. If these contracts were cancelled at December 31, 2015 and 2014, the Organization would have owed $109,521 and $61,965, respectively, to the Pittsburgh, Minneapolis, and Anaheim hotels.

NOTE 10 – PENSION PLAN AND RETIREMENT CONTRIBUTIONS

The University of Illinois (the University) contributes to the State Universities Retirement System of Illinois (SURS), a cost-sharing multiple-employer, defined-benefit pension plan with a special funding situation whereby the State makes substantially all actuarially determined required contributions on behalf of the participating employers. SURS was established July 21, 1941, to provide retirement annuities and other benefits for staff members and employers of the State universities, certain affiliated organization, and certain other State educational and scientific agencies and for survivors, dependents and other beneficiaries of such employees. SURS is considered a component unit of the State of Illinois' financial reporting entity, and is included in the State's financial reports as a pension trust fund. SURS is governed by Section 5/15, Chapter 40, of the *Illinois Compiled Statutes*. SURS issues a publically available financial report and required supplementary information. That report may be obtained by accessing the Web site at www.SURS.org or by calling 1-800-275-7877.

SURS federal tax identification number is 37-6006008. The plan was underfunded by $20,110,500,000 as of June 30, 2013, the most recent actuarial valuation date, which includes actuarial value of assets totaling $14,262,600,000 and accrued actuarial liabilities totaling $34,373,100,000. The plan was underfunded by $21,584,800,000 as of June 30, 2014, the most recent actuarial valuation date, which includes actuarial value of assets totaling $12,788,300,000 and accrued actuarial liabilities totaling $34,373,100,000.

While the Organization reports under Financial Accounting Standards Board, the University reports under Governmental Accounting Standards Board (GASB). GASB Statement No. 68, *Accounting and Financial Reporting for Pensions*, issued June 2012, will be effective for the University beginning with its year ending June 30, 2015. This statement requires governments providing defined benefit pensions to recognize their long-term obligation for pension benefits as a liability for the first time, and to more comprehensively and comparably measure the annual costs of pension benefits. This Statement also enhances accountability and transparency through revised and new note disclosures and required supplementary information. As the Organization participates in SURS through the University, this standard may impact the Organization's financial statements in addition to the University. Management is in the process of evaluating the potential impact this standard will have on its financial statements.

Eligible employees must participate upon initial employment. Employees are ineligible to participate if (a) employed after having attained age 68; (b) employed less than 50% of full time; or (c) employed less than full time and attending classes with an employer.

Plan members are required to contribute 8.0% of their annual covered salary and substantially all employer contributions are made by the State on behalf of the individual employers at an actuarially determined rate. The Organization's current employer rate is 12.69% of annual covered payroll. The contribution requirements of plan members and employers are established and may be amended by the Illinois General Assembly. During the years ended December 31, 2015, 2014 and 2013, the Organization reimbursed the University of Illinois $13,231, $14,743, and $16,436, respectively, for the employer's payments to SURS for eligible employees.

Employers hired prior to April 1, 1986 are exempt from contributions required under the Federal Insurance Contribution Act. Employees hired after March 31, 1986 are required to contribute 1.45% of their gross salary for Medicare. The Organization is required to match this contribution. As of December 31, 2015 and 2014, all employees were required to contribute 1.45% of their gross salary for Medicare.

Employees may also elect to participate in certain tax-sheltered retirement plans. These voluntary plans permit employees to designate a part of their earnings into tax-sheltered investments and thus defer federal and state income taxes on their contributions and accumulated earnings under the plans. Participation and the level of employee contributions are voluntary. The Organization is not required to make contributions to these plans.

NOTE 11 – POST-EMPLOYMENT BENEFITS

The State of Illinois (the State) provides health, dental, vision, and life insurance benefits for retirees and their dependents in a program administered by the Department of Central Management Services. Substantially all State employees become eligible for post-employment benefits if they eventually become annuitants of one of the State sponsored pension plans. Health, dental, and vision benefits include basic benefits for annuitants and dependents under the State's self-insurance plan and insurance contracts currently in force. Annuitants may be required to contribute toward health, dental, and vision benefits with the amount based on factors such as date of retirement, years of credited service with the State, whether the annuitant is covered by Medicare, and whether the annuitant has chosen a managed health care plan. Annuitants who retired prior to January 1, 1998, and who are vested in the SURS do not contribute toward health, dental, and vision benefits. For annuitants who retired on or after January 1, 1998, the annuitant's contribution amount is reduced five percent for each year of credited service with the State; therefore, those annuitants with twenty or more years of credited service do not have to contribute toward health, dental, and vision benefits. Annuitants also receive life insurance coverage equal to the annual salary of the last day of employment until age 60, at which time the benefit becomes $5,000.

The State pays the SURS' portion of employer costs for the benefits provided. The total cost of the State's portion of health, dental, vision, and life insurance benefits of all members, including postemployment health, dental, vision, and life insurance benefits, is recognized as an expenditure by the State in the Illinois Comprehensive Annual Financial Report. The State finances the costs on a pay-as-you-go basis. The total costs incurred for health, dental, vision, and life insurance benefits are not separated by department or component unit for annuitants and their dependents nor active employees and their dependents.

A summary of post-employment benefit provisions, changes in benefit provisions, employee eligibility requirements including eligibility for vesting, and the authority under which benefit provisions are established are included as an integral part of the financial statements of the Department of Central Management Services. A copy of the financial statements of the Department of Central Management Services may be obtained by writing to the Department of Central Management Services, Stratton Building, Room 720, Springfield, Illinois, 62706-4100.

LABOR AND EMPLOYMENT RELATIONS ASSOCIATION
SCHEDULE OF FUNCTIONAL EXPENSES
FOR THE YEAR ENDED DECEMBER 31, 2015

		MEETINGS			PUBLICATIONS						SUPPORTING SERVICES		
	General	Annual Conference	ASSA Meeting	Other Meetings	Directory	Newsletter	Annual Proceedings	Perspectives	Research Volume	Other Publications	Management & General	Membership Development	Totals
Compensation	$ 96,665	$ -	$ -	$ -	$ -	$ -	$ -	$ -	$ -	$ -	$ 17,922	$ -	$ 114,587
Payroll Taxes & Fringes	38,551	-	-	-	-	-	-	-	-	-	7,113	-	45,664
Contract Labor	31,021	-	-	-	-	-	-	-	-	-	-	-	31,021
Depreciation	-	-	-	-	-	-	-	-	-	-	739	-	739
Loss on Disposal of Fixed Assets	-	-	-	-	-	-	-	-	-	-	-	-	-
Insurance	-	-	-	-	-	-	-	-	-	-	3,843	-	3,843
Bank Charges	-	-	-	-	-	-	-	-	-	-	1,851	-	1,851
Promotion	-	2,906	149	-	-	-	-	-	-	-	-	3,397	6,452
Fulfillment	-	-	-	-	-	-	-	-	-	-	7,664	-	7,664
Postage and Freight	-	300	-	-	-	-	-	2,214	1,263	-	1,418	-	5,195
Accounting/Auditing	-	-	-	-	-	-	-	-	-	-	4,850	-	4,850
Printing and Production	-	2,490	139	-	-	-	-	7,505	7,338	43	-	-	17,515
Services	-	-	-	-	-	-	-	12,221	9,307	-	-	-	21,528
Other Publication Costs	-	-	-	-	-	-	-	1,390	-	323	-	-	1,713
Meals/Receptions	-	66,513	3,464	-	-	-	-	-	-	-	-	-	69,977
Travel	-	4,692	976	-	-	-	-	-	-	-	-	-	5,668
Other Meeting Expenses	-	7,269	410	11	-	-	-	-	-	-	-	-	7,690
Computer Supplies/Services	-	490	-	-	-	-	-	-	-	-	2,861	-	3,351
Website	-	-	-	-	-	-	-	-	-	-	1,900	-	1,900
Office Supplies	-	-	-	-	-	-	-	-	-	-	77	-	77
Student & Member Awards	-	-	-	-	-	-	-	-	-	-	-	3,042	3,042
Fundraising Expense	-	-	-	-	-	-	-	-	-	-	-	-	-
Duplicating Expense	-	-	-	-	-	-	-	-	-	-	315	-	315
Telephone & Fax	-	-	-	-	-	-	-	-	-	-	383	-	383
Chapter Expenses	-	-	-	-	-	-	-	-	-	-	-	-	-
Dues	-	-	-	-	-	-	-	-	-	-	-	-	-
Other Committee Expenses	-	-	-	-	-	-	-	-	-	-	-	273	273
Miscellaneous Office	-	-	-	-	-	-	-	-	-	-	43	-	43
	$ 166,237	$ 84,660	$ 5,138	$ 11	$ -	$ -	$ -	$ 23,330	$ 17,908	$ 366	$ 50,979	$ 6,712	$ 355,341

LABOR AND EMPLOYMENT RELATIONS ASSOCIATION
SCHEDULE OF FUNCTIONAL EXPENSES
FOR THE YEAR ENDED DECEMBER 31, 2014

		MEETINGS			PUBLICATIONS						SUPPORTING SERVICES		
	General	Annual Conference	ASSA Meeting	Other Meetings	Directory	Newsletter	Annual Proceedings	Perspectives	Research Volume	Other Publications	Management & General	Membership Development	Totals
Compensation	$ 112,163	$ -	$ -	$ -	$ -	$ -	$ -	$ -	$ -	$ -	$ 16,173	$ -	$ 128,336
Payroll Taxes & Fringes	45,087	-	-	-	-	-	-	-	-	-	6,706	-	51,793
Contract Labor	29,280	-	-	-	-	-	-	-	-	-	-	-	29,280
Depreciation	-	-	-	-	-	-	-	-	-	-	1,080	-	1,080
Loss on Disposal of Fixed Assets	-	-	-	-	-	-	-	-	-	-	445	-	445
Insurance	-	-	-	-	-	-	-	-	-	-	3,773	-	3,773
Bank Charges	-	-	-	-	-	-	-	-	-	-	2,731	-	2,731
Promotion	-	4,256	139	-	-	-	-	-	-	-	-	3,252	7,647
Fulfillment	-	-	-	-	-	-	-	-	-	-	20,240	-	20,240
Postage and Freight	-	707	52	-	-	-	-	1,403	1,110	-	228	-	3,500
Accounting/Auditing	-	-	-	-	-	-	-	-	-	-	4,800	-	4,800
Printing and Production	-	1,841	-	-	-	-	-	12,732	4,400	68	-	-	19,041
Services	-	-	-	-	-	-	487	12,737	6,715	-	-	-	19,939
Other Publication Costs	-	-	-	-	875	375	500	139	-	-	-	-	1,889
Meals/Receptions	-	47,666	-	-	-	-	-	-	-	-	-	-	47,666
Travel	-	3,602	1,889	611	-	-	-	-	-	-	-	-	6,102
Other Meeting Expenses	-	15,171	479	150	-	-	-	-	-	-	-	-	15,800
Computer Supplies/Services	-	2,405	-	-	-	-	-	-	-	-	439	-	2,844
Website	-	-	-	-	-	-	-	-	-	-	18,601	-	18,601
Office Supplies	991	-	-	-	-	-	-	-	-	-	1,263	-	2,254
Student & Member Awards	-	-	-	-	-	-	-	-	-	-	-	5,985	5,985
Fundraising Expense	-	-	-	-	-	-	-	-	-	-	-	994	994
Duplicating Expense	-	-	-	-	-	-	-	-	-	-	367	-	367
Telephone & Fax	-	-	-	-	-	-	-	-	-	-	438	-	438
Chapter Expenses	-	-	-	-	-	-	-	-	-	-	-	349	349
Dues	-	-	-	-	-	-	-	-	-	-	641	-	641
Other Committee Expenses	-	-	-	-	-	-	-	-	-	-	-	111	111
Miscellaneous Office	-	-	-	-	-	-	-	-	-	-	2,789	-	2,789
	$ 187,521	$ 75,648	$ 2,559	$ 761	$ 875	$ 375	$ 987	$ 27,011	$ 12,225	$ 68	$ 80,714	$ 10,691	$ 399,435

Also in the
Labor and Employment Relations Association Series

The Evolving Health Care Landscape: How Employees, Organizations, and Institutions are Adapting and Innovating
Edited by Ariel C. Avgar and Timothy J. Vogus.. $29.95

Inequality, Uncertainty, and Opportunity: The Varied and Growing Role of Finance in Labor Relations
Edited by Christian E. Weller.. ...$29.95

The Disunited States of America: Employment Relations Systems in Conflict
Edited by David Jacobs and Peggy Kahn...$29.95

Collective Bargaining Under Duress: Case Studies of Major U.S. Industries
Edited by Howard Stranger, Ann C. Frost, and Paul F. Clark...$29.95

Public Jobs and Political Agendas: The Public Sector in an Era of Economic Stress
Edited by Daniel J. B. Mitchell..$29.95

Employee Ownership and Shared Capitalism: New Direction in Research
Edited by Edward J. Carberry...$29.95

Transforming the U.S. Workforce Development System: Lessons from Research and Practice
Edited by David Finegold, Mary Gatta, Hal Salzman, and Susan J. Schurman..........................$29.95

Human Rights in Labor and Employment Relations: International and Domestic Perspectives
Edited by James A. Gross and Lance Compa $29.95

The Gloves-off Economy: Workplace Standards at the Bottom of America's Labor Market
Edited by Annette Bernhardt, Heather Boushey, Laura Dresser, and Chris Tilly...................................$29.95

Employee Pensions: Policies, Problems, and Possibilities
Edited by Teresa Ghilarducci and Christian E. Weller $29.95

Book orders and request for prices and order forms for the above LERA publications should be sent to:

ILR Press, Cornell Univ., P.O. Box 6525, Ithaca, NY 14851
Phone 607-277-2211, Fax 607-277-6292
Web www.cornellpress.cornell.edu
Additional volumes are available through ILR Press.

Information about ordering printed copies of the *Proceedings of the LERA Annual Meeting* can be found online at https://lera.memberclicks.net/publications. Orders for printed copies of *Perspectives on Work* and other LERA publications should be sent to the LERA Office:

LERA, 121 LER Building, 504 E. Armory Ave., MC-504, Champaign, IL 61820
Phone 217-333-0072, Fax 217-265-5130
Email LERAoffice@illinois.edu, Web leraweb.org

Alphabetical List of Authors

You are invited to become a member of

THE LABOR AND EMPLOYMENT RELATIONS ASSOCIATION

The Labor and Employment Relations Association (LERA) was founded in 1947 by a group who felt that the growing field of industrial relations required an association in which professionally minded people from different organizations could meet. It was intended to enable all who were professionally interested in industrial relations to become better acquainted and to keep up to date with the practices and ideas at work in the field. To our knowledge there is no other organization that affords the multiparty exchange of ideas we have experienced over the years–a unique and valuable forum. After 69 years, both our academic and practitioners agree with the conviction of the founders that the encouragement, reporting, and critical discussion of research are essential if our professional field is to advance.

Our local and national membership of 5,000 includes representatives of management, unions, government; practitioners in consulting, arbitration, mediation, and law; and scholars and teachers representing many disciplines in colleges and universities in the United States and Canada, as well as abroad. Libraries and institutions interested in the publications of the Association are also invited to become subscribing members and thereby receive the same member publications. Organizational memberships in the Association are also available.

Membership dues cover publications for 12 months and entitle members to the electronic newsletter, membership directory and proceedings of the annual meeting online, and printed copies of the annual research volume, and the magazine, *Perspectives on Work* (volumes also available online). Additional online resources include Chapter Profiles and an IR/HR Degree Program Listing for the United States, Canada, and Australia.

Dues for the 2017 calendar year are listed below. International memberships require an additional $20.00 per year to cover postage costs.

Regular Member	$199.00
Contributing Member	$310.00
Emeritus Member (10 years+ membership and no longer employed)	$110.00
Student Member (full-time, limited to 4 consecutive years)	$30.00
Library Subscriber	$300.00

Tax-deductible financial contributions to the Association to support its educational activities are always welcome. You can make an online contribution at our website, leraweb.org, and choosing "Donate". If you are not already a member, we invite you to join the LERA by sending your membership application and dues payment. You can also join online at the LERA website, leraweb.org by choosing "Membership", or request an invoice. Inquiries regarding membership, meetings, and publications should be addressed to the LERA office.

LABOR AND EMPLOYMENT RELATIONS ASSOCIATION
University of Illinois at Urbana-Champaign
121 LER Building, 504 E. Armory Avenue
Champaign, IL 61820 USA
Phone: 217-333-0072 Fax: 217-265-5130
Email: LERAoffice@illinois.edu Web: leraweb.org

Also in the
Labor and Employment Relations Association Series

The Evolving Health Care Landscape: How Employees, Organizations, and Institutions are Adapting and Innovating
Edited by Ariel C. Avgar and Timothy J. Vogus... $29.95

Inequality, Uncertainty, and Opportunity: The Varied and Growing Role of Finance in Labor Relations
Edited by Christian E. Weller.. ...$29.95

The Disunited States of America: Employment Relations Systems in Conflict
Edited by David Jacobs and Peggy Kahn.. $29.95

Collective Bargaining Under Duress: Case Studies of Major U.S. Industries
Edited by Howard Stranger, Ann C. Frost, and Paul F. Clark.. $29.95

Public Jobs and Political Agendas: The Public Sector in an Era of Economic Stress
Edited by Daniel J. B. Mitchell..$29.95

Employee Ownership and Shared Capitalism: New Direction in Research
Edited by Edward J. Carberry...$29.95

Transforming the U.S. Workforce Development System: Lessons from Research and Practice
Edited by David Finegold, Mary Gatta, Hal Salzman, and Susan J. Schurman.......................$29.95

Human Rights in Labor and Employment Relations: International and Domestic Perspectives
Edited by James A. Gross and Lance Compa $29.95

The Gloves-off Economy: Workplace Standards at the Bottom of America's Labor Market
Edited by Annette Bernhardt, Heather Boushey, Laura Dresser, and Chris Tilly................ $29.95

Employee Pensions: Policies, Problems, and Possibilities
Edited by Teresa Ghilarducci and Christian E. Weller $29.95

Book orders and request for prices and order forms for the above LERA publications should be sent to:

ILR Press, Cornell Univ., P.O. Box 6525, Ithaca, NY 14851
Phone 607-277-2211, Fax 607-277-6292
Web www.cornellpress.cornell.edu
Additional volumes are available through ILR Press.

Information about ordering printed copies of the *Proceedings of the LERA Annual Meeting* can be found online at https://lera.memberclicks.net/publications. Orders for printed copies of *Perspectives on Work* and other LERA publications should be sent to the LERA Office:

LERA, 121 LER Building, 504 E. Armory Ave., MC-504, Champaign, IL 61820
Phone 217-333-0072, Fax 217-265-5130
Email LERAoffice@illinois.edu, Web leraweb.org